YOU
ENVI

YOU AND THE ENVIRONMENT

Edited by KATIE McBRATNEY

Published by Consumers' Association
and Hodder & Stoughton

CONTRIBUTORS   David Attwood, Jackie Bennett, Peter Burgess,
Jane Chumbley, David Cope, Barbara Harvey, Kim Healy, Sian Morrissey,
John Reynolds, Simon Richmond, Isabelle Risner, Tess Sullivan,
Ronald Toms, Stella Yarrow

Which? Books are commissioned and researched by
The Association for Consumer Research
and published by Consumers' Association
2 Marylebone Road, London NW1 4DX, and
Hodder & Stoughton, 47 Bedford Square,
London WC1B 3DP

Typographic design by Paul Saunders
Cover artwork by John Holder

First edition 1990

British Library Cataloguing in Publication Data

You and the environment
   1. Great Britain. Environmentally safe household products
   I. Consumers' Association
   648

ISBN 0 340 52797 8

Typeset by Litho Link Ltd, Welshpool, Powys
Printed and bound in Great Britain by Collins, Glasgow

CONTENTS

PREFACE

The state of the environment, whether on a global or a local level, is now such that everyone must participate if we are to halt and reverse its decline. Consumers have a vital role to play in that process, since by orientating their choice in favour of products whose production has caused less harm to the environment and which generate less waste and fewer harmful substances they can influence the producers and manufacturers towards environmentally friendly products. To enable consumers to make these choices in a properly informed way, a system of ecological labelling is clearly necessary. Certain Member States of the European Community already have such systems in place, and others are actively considering introducing them. After the completion of the Internal Market in 1992, consumers will have a Community-wide role to play, and a European labelling scheme is vital in the Single Market. The Commission of the European Communities is presenting proposals for such a scheme which, if it is to be truly effective, must cover the environmental aspects of all the stages in a product's life, from its manufacture to its final disposal. Armed with this knowledge, consumers will be able to make informed choices, and each individual and apparently ineffective act will become a vital component of a larger action in favour of the environment.

The Honourable Carlo Ripa di Meana,
Commissioner of the European Communities
with responsibility for the environment

INTRODUCTION

Every day we are bombarded with information about the environment: about how our weather's changing and the disastrous consequences this could bring; how we're depleting the Earth's natural resources; how we're being wasteful, polluting and over-consuming. And it seems that everyone's getting involved in telling us what to do, from politicians to scientists and pressure groups. The world is beginning to pull environmental issues to the forefront of its mind, and it looks as if 'green' is here to stay.

So what can you do to sort fact from fiction in the welter of well-meaning advice about 'going green'? It's not easy, and some of the advice that's bandied around is distinctly off-putting. But *You and the Environment* can help you decide what shade of green you want to be. Why should you have to take time to sort out your rubbish for recycling, or spend money on getting your car modified (or even give it up), or change your favourite brand of soap powder? The answer is that you shouldn't, unless you are convinced that the compromises you have to make are worth any help you might be giving the environment.

You and the Environment examines in a balanced way the issues which everyone's worrying about – how the problems have come about, how they'll affect you and what you can do to help. The compromises you might have to make are pointed out too, so that you can decide for yourself whether you can put up with them.

There's lots of practical information for you to draw on, based on research carried out by *Which?*. *You and the Environment* will help you find kitchen appliances that use less water and power (and tell you how much they cost); advise you on using different gardening techniques to avoid chemicals; show you how you can help save the rainforests at a distance of thousands of miles; and how to see through the 'green' claims on everyday products that you buy. It will show you how to save energy in the home (saving on bills, too); how to use waste-recycling schemes; what to look for if you want organic food; and how to shop with the environment in mind. And in case you're inspired to tackle the larger issues, there's a guide to the main pressure groups working to save the environment.

You may not be able to change the world in one fell swoop, but at least you can decide for yourself what measures are most worthwhile – and understand the drawbacks involved.

=== 1 ===

WORDS OF WARMING – THE GREENHOUSE EFFECT

Is the world really on course for climatic catastrophe? Can you help to stop the greenhouse effect which, scientists say, might cause flooding and droughts, disruption of weather patterns and have serious effects on crops? There may seem to be little you can do about such a huge problem, but an optimistic attitude is that every little helps.

Many of us basking in the heatwave of summer 1989 must have wondered if global warming had arrived ahead of schedule. But slow-moving high-pressure areas and other natural weather conditions can conspire to give Britain a good summer. Climatologists are more concerned about a more gradual rise in global temperatures over a longer period. But it is true that, taking a global average, the six hottest years this century were all in the 1980s.

Some people probably think that global warming is no bad thing if it gives us hotter weather. In fact, it's by no means certain that global warming will turn this country into a sun-kissed paradise; Britain may get warmer, but it could well get wetter too. The storms of winter 1989–90, though exceptional, came about in a conventional way – a deep depression caused by cold polar air clashing with warm air from the tropics. On the other hand, frequent bouts of extreme weather might herald a period of significant climate change.

The greenhouse effect isn't as simple as an unwanted phenomenon which we're causing by pollution. For one thing, we need the greenhouse effect – life on Earth simply wouldn't exist without it.

How the greenhouse effect works

The Earth's atmosphere allows much of the sun's radiation to pass through, warming up both the atmosphere and the land and sea beneath.

The Earth gives off heat in return – infra-red radiation – which cannot pass back into space so easily: it's absorbed by the atmosphere and some is re-emitted back to Earth.

Our atmospheric blanket of gases keeps us on average about 30°C (54°F) warmer than we'd otherwise be: the moon, for example, though it's at roughly the same distance from the sun as we are, has an average temperature of about −18°C (−0.4°F) with violent extremes of heat and cold by day and night.

Carbon dioxide (CO_2) and water vapour have always played a part in this natural process. But now it seems that industrial development, intensive agriculture, burning of fossil fuels (coal, oil and gas) and deforestation are increasing the concentration of CO_2 and other gases in the atmosphere and intensifying the greenhouse effect. It's the *additional* global warming this would cause which has convinced many scientists of the need for urgent action.

The big heat

Climatic records show that the world is getting warmer: there's been a gradual temperature increase since the middle of the last century. But what *isn't* so clear is how much of this warming can be blamed on increasing greenhouse-gas concentrations; the warming since the middle of the nineteenth century has actually been smaller than most current climatic models would predict from the way CO_2 has increased.

The Earth has certainly cooled and warmed before as ice ages have come and gone. Some short-term global warming is probably caused by naturally occurring patterns of wind and ocean currents. It's also thought that the heat given off by big cities may affect temperature records locally. However, widely accepted estimates do suggest that by the year 2030, greenhouse-gas levels will be twice what they were before the Industrial Revolution. This could eventually lead to a warming of between 1.5 and 4.5°C (3 and 8°F).

This increase may not sound very much until you realise that the world is only 4°C (7°F) warmer now than it was during the last ice age 18,000 years ago.

The effects of a rise in temperature of this extent could be dramatic, resulting in:

- an increase in sea-levels caused by thermal expansion and melting ice. This could cause flooding of low-lying countries like Bangladesh and The Netherlands, while in Britain parts of East Anglia and the Thames Estuary could be at risk unless new defences are built
- regional shifts in climate and disruption of weather patterns, with the likelihood of more frequent floods and droughts
- serious effects on crops in some countries. In others, higher average temperatures and rainfall could increase yields (though milder winters could also make pest control more of a problem).

The greenhouse cocktail

The pie chart on p. 14 shows how much each gas contributes to the greenhouse effect.

Carbon dioxide 50 per cent

The oceans absorb billions of tonnes of CO_2, as do trees and plants, which need it to live. But the rate at which we're burning fossil fuels – coal, oil and gas – pumps more CO_2 into the atmosphere than these natural processes can cope with.

- Generating one unit of electricity in a coal-fired power-station produces about one kilogramme (2.2 pounds) of CO_2.

13

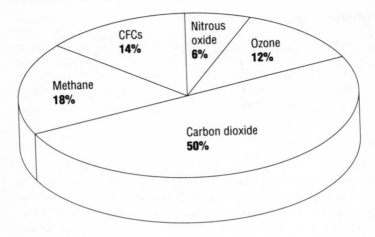

- Using one litre of petrol produces about 2.2 kilogrammes (4.8 pounds) of CO_2. Fuel burnt for transport produces a quarter of the CO_2 produced each year.
- Cutting down forests adds to CO_2 levels if the timber is burnt, and leaves less vegetation to absorb CO_2.
- Tests on air bubbles trapped in polar ice-caps suggest that atmospheric CO_2 levels remained fairly constant for thousands of years until the Industrial Revolution. Since then they've increased by 25 per cent – 10 per cent since the 1950s.

CFCs 14 per cent

Most people know that CFCs can damage the ozone layer (see Chapter 2), but CFCs are also extremely powerful greenhouse gases. Though they're being used less in aerosols and packaging, they're still widely used as refrigerants and industrial solvents – and, once released, remain in the atmosphere for many years. And some substitutes which claim to be 'ozone friendly' are not 'greenhouse friendly'.

14

Methane 18 per cent

Methane is produced by landfill rubbish tips, rotting wood, livestock, and even paddy fields. Natural gas before it's burnt is mostly methane, and some of this may escape during extraction and distribution. Methane concentrations in the atmosphere have doubled over the last 200 years.

Ozone 12 per cent

Ozone is good news in the upper atmosphere because it protects us from the sun's more damaging ultraviolet radiation (which can increase the risk of skin cancer). But nearer ground-level – where it's increasing due to the effect of sunlight on atmospheric pollution – ozone is another greenhouse gas.

Nitrous oxide 6 per cent

Nitrous oxide is thought to be increasing in the atmosphere for reasons that are not fully understood. The biggest known source is the growing use of agricultural fertilisers.

What should be done

Many scientists believe that if nothing is done, climatic change is only a matter of time. And the time-lag between cause and effect means that some global warming is probably inevitable.

Planning to minimise the impact of inevitable warming, for example by strengthening sea defences, is obviously important, but it's even more vital to minimise further warming. This, above all, means taking steps to save energy internationally without unreasonably burdening developing countries. An internal Department of Energy document in 1989 predicted that UK emissions of CO_2 would rise by 37 per cent by 2005 (though some recent studies have revised this

15

forecast downwards). The government has agreed that CO_2 levels should be stabilised, but some government departments say that the scientific evidence isn't yet strong enough to justify action which might have a draconian effect on our industrial competitiveness and standard of living. But West Germany and Japan, for example, both manage to combine energy efficiency with a high standard of living. One study published in 1989 showed that energy consumption from fossil fuels could be cut by nearly a quarter over the next 16 years and still allow the British economy to grow.

Market forces can encourage energy saving (especially if prices rise dramatically, as happened in the 1970s oil crisis), but they do not reflect the cost of environmental damage. There's something to be said for energy sources being priced to reflect their relative pollution potential, though this is difficult to do if one environmental problem has to be traded against another. For example, nuclear power-stations produce much less CO_2 than coal or oil ones do, but radiation hazards and waste disposal problems have to be taken into account. And coal-fired power-stations can have 'scrubbers' fitted to remove sulphur dioxide (which causes acid rain), but this may *increase* CO_2 output by reducing efficiency and requiring more fuel to be burnt.

Consumers' Association would like to see from government:

- a government-backed energy-saving programme as a priority. Cuts imposed on the budget of the Energy Efficiency Office should be restored
- the requirement that British Gas and the privatised electricity supply industry adopt 'least-cost planning' policies; these give equal consideration to investment in increasing energy efficiency (controlling demand) as well as in building new power-stations (increasing supply)
- readily available information on the energy efficiency of buildings. In Denmark, for example, all homes when they're sold must be given a certificate showing their energy efficiency. The latest UK Building Regulations do

include more stringent energy efficiency specifications for *new* buildings but not for old

- a comprehensive review of subsidy and investment strategies for public transport and road-building to take energy conservation into account
- tax incentives to encourage the use of fuel-efficient, low-pollution vehicles
- more investment and research into renewable energy resources such as solar, wind and wave power. 'Combined heat and power' – putting waste heat from power-stations to good use – is another candidate
- the re-introduction of incentives to consumers to insulate their homes
- further encouragement to recycle waste which otherwise adds to methane levels
- controls on the use of nitrogen-based fertilisers.

We'd like to see from industry:

- energy labelling of appliances so that consumers can readily identify which ones use least electricity or gas. Manufacturers seem reluctant to do this voluntarily, but such appliances would be cheaper to run as well as being more environmentally friendly. Legislation has been introduced in the USA which will outlaw fridges, freezers and air-conditioners that don't meet energy-efficiency criteria. The most efficient washing machine tested by *Which?* costs about £14 a year to run on a standard usage cycle, and the least efficient about £29 a year. If they both ran off electricity from a coal-fired power-station over a year, using the least efficient would produce over 200 kilogrammes (440 pounds) more CO_2. How much CO_2 an appliance produces depends not only on its power rating (in watts) but on how long it's in use. Washing machines, tumble driers and irons have a much higher power rating than fridges and TVs, but they're not in use for as long
- more thought given to the 'total energy' required to manufacture and ultimately dispose of an appliance, as well as to run it.

=== 2 ===

SEE OFF THE CFCs

Chlorofluorocarbons (CFCs) are used in the manufacture of some aerosols, plastic foams and refrigeration equipment, and as industrial cleaners and solvents. It's now widely accepted that CFCs affect the Earth's atmosphere in a way that could cause serious health risks and changes in climate.

The ozone layer

The ozone layer is vital to life on Earth; it shields us from the most harmful types of ultraviolet (UV) light given off by the sun. If more of this UV light gets through, the Earth will feel the effects – the number of skin cancer cases will increase, and there's strong evidence that increased UV exposure would harm crops, with serious consequences for the world's food supplies.

CFCs released into the atmosphere damage the ozone layer, reducing its effectiveness in filtering out some of the UV light. What happens is that CFCs from, say, aerosol use or industry make their way up to the stratosphere, above the Earth's surface (see diagram on page 19). Under the influence of strong sunlight, CFCs break down, producing chlorine, which attacks ozone gas, converting it to oxygen. The chlorine isn't destroyed in the process, so a little chlorine can continue doing a lot of damage, steadily depleting the ozone concentration.

There's another reason to worry about CFCs: their accumulation in the lower atmosphere – the troposphere – along

Where the ozone layer is

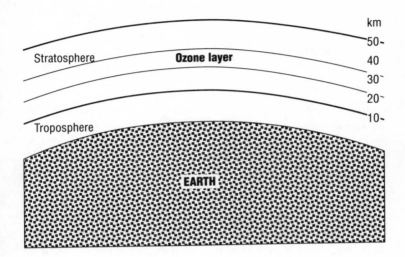

with other gases we produce, like carbon dioxide, is thought to be causing the Earth's average temperature to rise slightly (see Chapter 1).

The chemicals in question

Chlorofluorocarbons were first developed in the USA in the 1930s by Thomas Midgley (who was also responsible for first adding lead to petrol). Their main use then was refrigerants – chemicals used in fridges and air-conditioning plant. Since the 1950s they've come to be used for lots of other things; the pie chart on page 20 shows how the CFC usage slices up.

The CFC family of chemicals has several worthwhile properties – the chemicals are odourless, non-flammable and non-toxic (and because CFCs are so useful, chemical companies are working to develop new ones which don't

19

**What CFCs are used for
(1989 figures for the UK)**

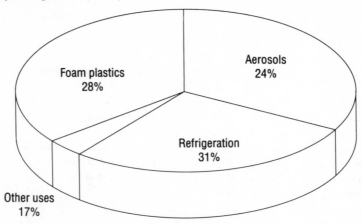

damage ozone). Different CFCs are identified by numbers: those most widely used up to the present have been CFC11 and CFC12. These are also two of the CFCs most harmful to the ozone layer – they have a high ozone depletion potential.

Although people often talk about 'CFCs' in general, the less harmful fluorocarbons are more correctly called HCFCs, and the *least* harmful ones are called HFCs. In these alternative CFCs, the chlorine is partly or wholly replaced with hydrogen. HCFC22 is 20 times less damaging than CFC11 or 12; HFC134a (a new refrigerant) doesn't damage ozone at all. Manufacturers of 'ozone friendly' products may have switched over to less harmful CFCs or got rid of them altogether. Unfortunately, such replacements (especially HFCs) may still contribute to the greenhouse effect.

The Table lists all the fluorocarbons and similar substances currently in use and shows what they're used for and how harmful they are to the environment.

Halons – chemicals with similar ozone-damaging properties but containing the element bromine instead of chlorine – are widely used in fire extinguishers.

FLUOROCARBONS AND RELATED SUBSTANCES

CFC	ODP	GWP	MP	Comments and uses
CFC11	100	100	Y	Foam, refrigeration, aerosols
CFC12	90–100	280–340	Y	Foam, refrigeration, aerosols
CFC113	80–90	130–140	Y	Dry-cleaning, solvent
CFC114	60–80	370–410	Y	Refrigeration, aerosols
CFC115	30–50	740–760	Y	Refrigeration
HCFC22	4–6	32–37	N	Refrigeration, aerosols
HCFC123	1.3–2.2	1.7–2.0	N	Foam, solvent#
HCFC124	1.6–2.4	9.2–10	N	Refrigeration#
HCFC225	1–5	3	N	Solvent
HCFC141b	7–11	8.4–9.7	N	Refrigeration, solvent#
HCFC142b	5–6	3.4–3.9	N	Foam*
HFC125	0	51–65	N	Refrigeration#
HFC134a	0	24–29	N	Refrigeration, aerosols
HFC143a	0	72–76	N	
HFC152a	0	2.6–3.3	N	*
Halon 1211	220–300	–	Y	Fire extinguishing
Halon 1301	780–1320	–	Y	Fire extinguishing
Halon 2401	500–620	–	Y	Fire extinguishing
Carbon tet.	100–120	34–35	N	Solvent
Methyl chl.	10–16	2.2–2.6	N	Solvent, aerosols, adhesives

Explanations
ODP – the ozone depletion potential (CFC11 = 100)
GWP – the CFC's greenhouse warming potential (CFC11 = 100)
MP – whether the CFC is controlled by the 1987 Montreal Protocol
– toxicity testing not complete
* – is slightly or moderately flammable

Controlling the problem

The CFC/ozone-depletion theory was first put forward by two American scientists in 1974. It met with initial scepticism, but continued research since then, under the United Nations Environment Programme (UNEP), has led to its substantial acceptance. In 1984 a huge 'hole' in the ozone layer was discovered above the Antarctic. However, there's still some disagreement about how serious the problem is, and how factors other than CFCs may be involved.

The USA banned non-essential use of CFC-based aerosols in 1978 (CFCs are at present still essential for some medical purposes like asthma sprays). A few other countries introduced voluntary restrictions or labelling requirements, but nothing much happened until March 1985, when more than 20 countries signed the Vienna Convention for the Protection of the Ozone Layer. This was followed by the signing in September 1987 of the Montreal Protocol, which demands cuts in consumption and production of certain CFCs and halons. The Protocol formally came into force in January 1989, and required that by 1999 consumption of specified CFCs – shown in the Table – would be cut by 50 per cent from 1986 levels.

But because CFCs (and the chlorine produced when they break down) persist in the atmosphere, many scientists felt that the Protocol didn't go far enough; one study suggested that even if every country phased out CFCs immediately, it would still take until the year 2050 to get atmospheric chlorine levels back to what they were in 1985. And although under the terms of the Protocol consumption would be cut by 50 per cent, a complex set of rules (drawn up so that developing countries are not unreasonably penalised) meant the *production* might be cut by as little as 35 per cent over the same period.

In March 1989 reports from NASA and British scientists revealed that chemical processes similar to those which caused the Antarctic ozone 'hole' were taking place in the ozone layer above the Arctic. In May of the same year a

22

meeting of the Protocol countries acknowledged that the Protocol was inadequate, and called for an end to all CFC production and use by the year 2000.

A further meeting of the Protocol countries in London in June 1990 agreed to strengthen the Protocol, involving the elimination of CFC production by the year 2000, and the addition to the list of controlled substances of further fluorocarbons and other chemicals such as the industrial solvents carbon tetrachloride and methyl chloroform. They also agreed financial and technical assistance to less developed countries to help persuade countries such as China and India, which have not yet signed the Protocol, to do so. In the meantime the European Commission is considering an even tighter package of controls.

Uses of CFCS

Aerosols

An aerosol can contain a product (deodorant, for example) and a liquefied propellant gas under pressure. When you press the button, the pressure in the can forces out a fine spray of product mixed with propellant. The propellant evaporates, leaving the product behind.

Formerly, CFC11 and CFC12, two of the CFCs most harmful to the ozone layer, were widely used as propellants, but following pressure from consumers and environmental groups at least 90 per cent of UK aerosols no longer use them. Most use hydrocarbons (such as butane or propane) for domestic cleaning sprays. Personal-care products like hairspray or shaving foam tend to use either hydrocarbons or a CFC that's less harmful to ozone. There's also been a swing towards trigger sprays for cleaning products and pump sprays for toiletries, neither of which uses a propellant gas.

Many manufacturers are also using labels such as 'ozone friendly' or 'contains no propellant alleged to damage ozone' on products that do not (and in many cases never did) contain harmful CFCs (see Chapter 3).

Aerosols which still use CFCs are mostly industrial products, medical products like asthma inhalers and pain-relieving sprays. Some d-i-y insulating foams contain CFCs too, and some stain-removers contain methyl chloroform which, though not a CFC, also harms ozone.

Even before concern grew about CFCs, hydrocarbons were used as propellants in many aerosol products – partly because they're cheap. But they're far from ideal for the purpose: for one thing – unlike CFCs – they're highly flammable. Because the contents of an aerosol can are under pressure, there's a risk of explosion if it's punctured or overheated, and using hydrocarbons increases the hazard. So there's scope for the development of new kinds of aerosol, one of which is the 'bag-in-can' type, which can use compressed air or other harmless gases as a propellant.

'Essential' medical aerosols are exempt from Protocol control, but there's a need to tighten controls over what's defined as essential. Asthma sprays clearly are; sprays designed to relieve muscular pain – and which contain at least 90 per cent harmful CFCs – should not be classed as essential.

What you can do

- Avoid aerosol products that use CFCs – check that the label states that the product is CFC-free before you buy. And, of course, trigger sprays, pump sprays and non-spray alternatives (like roll-on deodorants, for example) are not only 'ozone friendlier' but may also be cheaper.
- Heed the warnings on aerosol cans; keep them away from cooker tops and sunny window-sills, and don't spray near a flame or hot electric element.
- Don't buy aerosol sprays to numb muscular pain. For minor muscular pain, try an ice-pack (or warmth) or a painkiller like paracetamol.
- Check the labels on stain-removers and spray glues to make sure they don't contain methyl chloroform (1, 1, 1

trichloroethane).

- There are alternative asthma inhalers which don't use CFCs – the dry powder type, for example. Unfortunately, this is more expensive and may not suit all users. If you use an inhaler and you'd like to try an alternative, ask your doctor's advice.

Domestic fridges and freezers

Refrigerators are a particularly urgent global concern because ownership in countries such as China and India – at present low – is set to increase rapidly.

A fridge or freezer uses CFC12 as a refrigerant in its cooling system. CFC11 is used as a 'blowing' agent to manufacture the insulating plastic foam fitted in the appliance's walls and doors. Until recently, the average fridge contained about 100 grammes (3.5 ounces) of CFC in its cooling system (equal to about one and a half old-style aerosols) and about five times as much in the insulation.

The CFC12 refrigerant is sealed in the fridge and won't normally escape until the appliance is dumped. A little CFC11 is released into the atmosphere when the insulating foam is manufactured, but most is released when it's broken up.

Less harmful refrigerants and blowing agents are being developed as fast as possible. They need to be fully tested to make sure they're safe and non-toxic, and appliances will need modification to suit them. It seems likely that the new ozone-friendly refrigerant HFC134a will be commercially available by 1991, though fridge manufacturers will not be able to use it for some time. In the meantime manufacturers and retailers are trying to tackle the problem by recycling and by using less CFCs in the first place. These are their options:

Recycling This is the best option for the refrigerant. Some retailers will collect your old appliance and recycle the refrigerant if you buy a new fridge or freezer from them. Most refrigerator companies are looking into ways of handling

CFCs safely during servicing or disposal. A collection and recycling scheme is also available from a few local councils in conjunction with the main CFC suppliers.

Less CFCs Recycling the CFCs in the insulation isn't really practicable at present, and using fewer CFCs (or reducing the amount released during manufacture) is the answer. Most appliance manufacturers have now reduced the foam CFC content by 50 per cent. Further reductions can be difficult because poorer insulation can lead to reduced capacity and/or increased energy consumption. Because the CFCs in the insulation are so difficult to recycle, there's an urgent need for alternative types of insulation to be developed. One manufacturer, Bosch, has recently announced the development of a completely CFC-free foam.

What you can do
- If you need a new fridge or freezer, buy from a retailer that will collect your old appliance and recycle the refrigerant. The more people ask for this service, the more widespread it's likely to become. Or check if there's a scheme provided by your local authority.
- Check that the appliance you choose has a reduced CFC content in the insulation, and – in years to come – an ozone-friendly refrigerant too.

Large-scale freezing and cooling

Supermarket freezers, air-conditioning in shops, offices and cars, and in blood banks, munitions stores and mortuaries, all use CFCs in their cooling systems. In fact, the air-conditioning plant in a factory or superstore may contain about three-quarters of a million times as much CFC as a domestic refrigerator. And to add to the problem, large amounts of CFCs may be released whenever such plant is installed, serviced or scrapped. The only good news is that a lot of the coolant used is HCFC22, which is less harmful than CFC11 and CFC12. Most large supermarket chains say their contractors recycle the CFCs when old refrigeration equipment is removed from stores. New freezers generally use HCFC22.

There's a Code of Practice covering the installation, servicing, use and repair of refrigeration and air-conditioning plant. It's up to the supermarkets and other large users of such equipment to specify that the Code is being adhered to. The development of safer CFCs is a longer-term priority. And in the UK's temperate climate there should be scope for designing public buildings to need less air-conditioning; this would reduce CFC usage and save significant amounts of energy into the bargain.

Plastic foams and packaging

CFCS – mainly CFC11 – are used as a foaming agent in the manufacture of many rigid and flexible plastic foams. This has been used for all sorts of things:

- food packaging – fast-food containers and supermarket produce trays (but not the white expanded polystyrene used for packaging things like electrical goods)
- fridge insulation – see above
- furniture, car seats, pillows, even shoe soles
- building materials – rigid panels for walls and roofing. CFC-blown foam is widely used for these purposes because it's water-resistant and a good insulator – increasingly important with greater attention being paid to energy conservation in Building Regulations. Some d-i-y foams used for things like sealing around window-frames also contain CFCS.

Most packaging suppliers no longer use harmful CFCS to produce their packaging. A hydrocarbon, such as pentane, or HCFC22 is most often used. Hydrocarbons are preferable, because HCFC22 does damage ozone to some extent.

Manufacturers of rigid foam building materials are progressively reducing the use of CFC11 and introducing alternatives, such as HCFC123. Flexible foam manufacturers are switching in the short term to methylene chloride, and they plan to use alternatives to CFCS when available. For insulation purposes, alternatives (such as expanded poly-styrene or mineral wool) already exist, and others are being developed, but they are often more expensive or less efficient in their other properties such as strength, heat insulation or fire-resistance. CFC recovery isn't generally practicable, so reducing the amount used, modifying processes to reduce escape during manufacture, or using alternative gases is necessary. It's possible for architects to speed up the process of change by careful specification of the materials to be used.

Dry-cleaning

One of the two solvents generally used in dry-cleaning equipment is CFC113, a CFC controlled by the Montreal Protocol. It's good for dealing with delicate fabrics. Most of the CFC is used again and again, though a small amount will escape to the atmosphere as clothes dry after cleaning.

Consumers' Association believes that dry-cleaners should use the alternative – perchlorethylene – wherever possible (it's less good for delicate fabrics, leather and suede) and keep CFC emissions to a minimum by good working practices.

Industrial solvents

CFC113 is also widely used by the electronics and engineering industry for the cleaning and degreasing of components. In Japan 40 per cent of that country's CFCs are used for this purpose. Once again, preventing unnecessary loss and the development of alternatives are the key. Recently, several large manufacturers of microchips promised to phase out CFC use by the year 2000. However, one of the alternatives to CFC113 which might be used is methyl chloroform. Although this has an ozone depletion potential eight times less than CFC113, it still has a significant effect, and it is to be controlled by the strengthening of the Montreal Protocol. It would be better for manufacturers to switch directly to an ozone-benign alternative.

Fire extinguishers

Halons are chemicals which damage the ozone layer in much the same way as CFCs, except that weight for weight they're even more destructive. Their main use is in fire extinguishing equipment. Car fire extinguishers may use halon – check the labelling before you buy. There are also large-scale systems designed to flood, say, a library or computer room with halon gas if fire breaks out. Halon is preferred to liquid or powder extinguishers for this kind of use because it causes little damage. Up to now this type has usually been tested by releasing large amounts of halon into the atmosphere.

Large halon systems should be restricted to situations where they really are essential; other types of extinguisher such as water, dry powder or carbon dioxide should be used wherever possible. And every attempt should be made to limit the amount of halon released during installation, testing and firefighter training.

What you can do
Considering that it would be used only in an emergency, you may feel that a halon extinguisher is justifiable in your car. Dry powder types are an alternative, though more messy when let off. Fire safety experts are not generally very keen on fire extinguishers for use in the home, because they may not be effective if not regularly maintained or if let off by an inexperienced user. But if you want one, look for a foam or dry powder type.

=3=

GREEN GROCERY

If you've decided you want to buy products that are less damaging to the environment than others, shopping for them isn't as simple as you might think. Many of the labels promoting the apparent 'environmental friendliness' of some products are at best confusing and at worst misleading. There are currently no guidelines governing how a product should be labelled in environmental terms, so manufacturers and retailers can use any wording they like.

A survey conducted by *Which?* showed that many people are confused by environmental labels and their status. Almost 2000 people were interviewed to find out how much notice they took of 'green' labels and what they understood by them. *Which?* discovered that:

- over half the people interviewed believed that a label shown to them saying 'environment friendly' implied some sort of official approval. When they were asked who they thought had approved it, the most popular guess was the government. The label shown to them had in fact been devised by the marketing department of a high street chemist, and the products which carried it did not have to meet any official environmental criteria
- of the people who correctly thought that goods carrying 'green' labels don't require official approval, more than four out of five thought they *should* require it. Nearly six out of ten of the people asked thought that the government would be the appropriate body to give such approval.

Which? also carried out a series of group discussions with people responsible for doing the shopping for their household. There was a general feeling in these discussions that some manufacturers were simply jumping on the bandwagon and using 'green' labels as a marketing ploy. When the group discussion panels looked at a range of 'green' labels, it emerged that people were confused about what individual terms meant, and were suspicious that the claims may have been unjustified.

So which sorts of labels are most likely to confuse and mislead you when you go shopping? They can be broadly split into five main groups:

Excessive claims

No manufactured product can fail to have some sort of negative impact on the environment. Labels which claim that a product is 'environmentally friendly' or 'green' are very misleading. For example, even if the trees used to make paper products like sanitary towels are from a properly managed forest, and even if the paper used in the product is unbleached or has been bleached using less polluting types of bleach, pulp and paper production are still highly energy-intensive processes – energy which has to be provided at the cost of the environment. Deodorant and hairspray manufacturers who market their products in aerosol containers and put on the container a label saying 'environmentally friendly' may avoid using the most harmful CFCs as a propellant, but the alternative propellant gases they use can still damage the atmosphere, and the aerosol containers themselves are not particularly 'environmentally friendly'. (See Chapter 2 for more about aerosols.)

Meaningless claims

Some products display labels saying that the product does not contain a certain ingredient or have a certain property that has been linked with some sort of environmental

31

damage, when products of that type would never contain that ingredient in the first place. It's misleading, for example, to put a 'no nitrates' label on a bathroom cleaner when you can't buy one that *does* contain nitrates. Similarly, a 'phosphate free' label on a bottle of washing-up liquid suggests that this product is environmentally better than other washing-up liquids which *do* contain phosphates – but *no* washing-up liquid on sale in the UK contains phosphates.

Claims which aren't explained

Some products can confuse consumers with technical or scientific terms which aren't explained properly. When the discussion groups were shown a label from a washing powder saying 'no phosphates, no NTA, no enzymes, no optical brighteners', most people didn't know what these terms meant. The term 'biodegradable' on the label of a washing-up liquid was not fully understood, nor was the meaning behind the words 'environmentally friendly pulp' on a packet of disposable nappies. None of these packets gave a proper explanation of what was meant by the claims, so it was difficult for shoppers to assess how important they were in environmental terms. It's confusing for information on packets to be so sparse that customers cannot make an informed choice about what they are buying.

Multiple claims

Where products are making essentially the same environmental claim, standard wording should be used to avoid unnecessary confusion. For example, now that many manufacturers and retailers use aerosols which don't contain the most harmful CFCs as a propellant, many different kinds of labels appear on aerosols with different forms of wording. Members of the discussion groups thought that an aerosol labelled 'ozone safe' was somehow 'greener' than one labelled 'ozone friendly'. In fact, both contained hydrocarbons as a propellant – so both were the same.

Unrealistic claims

You may come across labels on plastic containers stating that the container is recyclable. It is helpful for a manufacturer to point out what packaging is made of and to state whether it is recyclable. At present, though, plastic recycling schemes and collection points are few and far between. Consumers may be persuaded into buying a product partly because it has a 'recyclable' label on the plastic container, but when it comes to disposing of it there's no local recycling point, so it just joins the rest of the rubbish heap. (On the other hand, labelling containers in this way may help result in more recycling schemes being set up – but consumers should still be aware of the realities of the situation.)

What can be done?

Both consumers and manufacturers are confused by environmental labels – consumers because labels are often unclear and misleading, and manufacturers because they're not sure how best to word labels on containers. Fortunately, however, a possible end to this confusion is in sight, as the government is planning to set up an official environmental labelling scheme.

Under the scheme, products will be awarded an official environmental label after scrutiny by an independent panel of environment 'judges'. The panel is likely to include representatives from consumer groups, environmental groups, manufacturers and retailers. Manufacturers and retailers would submit their products for inclusion in the scheme. The products would then be awarded an official label if they are judged to have met certain environmental criteria for all stages in their life cycle – covering production, packaging, use and disposal: the so-called 'cradle-to-the-grave' approach.

Such a scheme is intended to operate throughout the European Community, although it's possible that the national scheme will be launched before a Europe-wide scheme. The government hopes to have the labelling scheme set up by 1991. Manufacturers won't be compelled to submit their products to the panel of environment judges, as the scheme is voluntary. This won't in itself prevent manufacturers from continuing to use their own labels, although the right to use an official label should be a strong marketing incentive. The government is, however, looking at the Trade Descriptions Act 1968 to see whether blatantly misleading claims like 'environmentally friendly' could be covered by the Act.

Official environment labelling schemes are nothing new. In West Germany, for example, the 'Blue Angel' scheme, run by the government, has been in operation since 1978. Thousands of products now carry the 'Blue Angel' showing they have met specific environmental criteria. Examples of products covered include those that are reusable or recyclable (such as glass bottles) and those that use up fewer natural resources in the manufacturing process (such as recycled plastic and paper products). The scheme is voluntary, and it hasn't prevented some manufacturers from devising their own environmental labels. Canada and Japan are two other countries that also already have environmental labelling schemes up and running.

Once an official labelling scheme is set up in the UK, it should be far easier for you to shop for 'greener' products. You'll know that products displaying an official environmental label will have been 'passed' by an independent panel, which has made sure that the product meets certain environmental criteria. The confusion surrounding many of the green labels currently on the shelves should soon be a thing of the past. In the meantime, try to go for products that seem to give reasonably clear information either on the packet or in separate leaflets explaining why an environmental label is shown. And treat with scepticism very general claims like 'environmentally friendly' or 'green'.

=4=

WASTE AWAY

Every year in the UK we throw out around 18 million tonnes of domestic waste, around a third of a tonne per person. A family of four – two adults and two children – puts out, on average, two full refuse sacks every week, for collection and disposal. Whatever grumbles we may have about the efficiency of the local waste-collection services, most of our rubbish is removed from our doorsteps every week. It's because it is collected and disposed of, out of sight, that it is easy for us to turn a blind eye to just how much rubbish we create.

Domestic waste disposal

The rubbish that we put in our bins doesn't just disappear, and there's an increasing problem of what to do with this 18 million tonnes of waste. Currently there are three main options for domestic waste disposal:

- burying it in landfill sites
- burning it in incinerators
- recycling the usable materials.

Landfill

It's estimated that over 90 per cent of the domestic waste created each year in the UK is currently put into landfill sites. These used to be fairly rudimentary, conveniently sited holes in the ground, where the waste was simply dumped and covered over. It was a cheap way to quickly dispose of waste, but poorly thought out landfill sites have been the cause of environmental concern.

One problem is how to control leachate, a liquid produced when the waste is broken down by bacteria and which mixes with water seeping through the site. On landfills which have an impermeable base and sides, the leachate is collected and disposed of. But on many sites, leachate sinks into the ground. Under natural processes of degradation, this leachate should be broken down, diluted and made harmless. However, some scientists are concerned that certain chemicals in the leachate may not break down safely and can contaminate the underground water system.

There are particular worries about co-disposal sites, where domestic and industrial wastes are disposed of together. Biological, chemical and physical reactions are supposed to make the mixture of chemicals harmless, but to do this successfully the types and volumes of waste deposited must be very carefully controlled. There's evidence this hasn't always been done.

Another major problem is landfill gas, a mixture of methane and carbon dioxide produced when the waste breaks down. It should be monitored and vented properly to stop it building up and possibly causing an explosion. A child was burnt to death in a landfill gas explosion in 1984. After another explosion in 1986 at Loscoe, in Derbyshire, a survey of landfills was carried out. Almost 1400 active or closed landfills were found in England and Wales which are a potential gas risk. More than half the sites are within 250 metres of houses or industry, but fewer than 30 per cent of these have any gas control measures. The survey did not cover landfills closed before 1978, which could also be a risk.

On modern landfill sites, the underlying rocks and soil are assessed for suitability, and the National Rivers Authority must be consulted about the risk of water pollution. But before laws regulating waste disposal were passed in the 1970s, chemical waste was sometimes dumped in an uncontrolled way. It's not known where all these old sites are, although they're a potential risk. Today, most domestic landfill sites are closely monitored. Rubbish is often covered over daily with clay and topsoil. Making these kinds of improvements to the way landfill sites are run, in order to eradicate the environmental problems associated with some old landfill sites, has made landfill a more expensive option for waste disposal than it used to be.

There is also the more fundamental question of whether we actually want to use land for dumping our waste, even if it can be done safely and efficiently. Because we will always create some waste that can't be recycled, safe landfilling is likely to play some part in waste disposal for the foreseeable future. But in the south-east of England, for example, suitable space for landfill sites is scarce. Currently, by burying over 90 per cent of all domestic waste in landfill sites, we are using up suitable sites very quickly – and mostly by filling them with unsorted domestic waste that contains valuable materials which could be recycled.

Incineration

The second option for waste disposal is to burn it. Municipal incinerators are owned and operated by local authorities; there are 35 to 40 in the UK. They burn domestic, commercial and non-hazardous industrial wastes. Some of these produce energy that is then available for district heating or commercial use. Burning waste to make energy sounds like a good solution, but incinerators are themselves a potential environmental hazard.

Incinerator chimneys emit smoke and gases containing a polluting mixture of chemicals and heavy metals. Municipal incinerators also emit dioxins. Some dioxins in sufficient quantities are toxic to humans, but the health risk of exposure to tiny amounts in the environment isn't really known.

Anti-pollution equipment installed at incinerators can reduce the emissions – 'scrubbers' to remove sulphur

37

dioxide, which can cause acid rain – but municipal incinerators in the UK were built 10 to 20 years ago and don't have the most up-to-date equipment. Strict new EC rules limiting incinerator emission-levels mean that most municipal incinerators are likely to close over the next few years. Local authorities will not be able to afford the equipment needed to meet the new standards.

Recycling

The greatest potential for reducing the amount of waste that needs to be incinerated or put in landfill sites lies in creating less rubbish in the first place and in recycling far more of our domestic waste.

The government has set a target for household waste recycling of 50 per cent of recyclable materials by the year 2000. Domestic waste contains a surprisingly high proportion of recyclable materials. But in the average dustbin, valuable raw materials like aluminium cans, glass bottles and paper are mixed up with leftover food, discarded packaging and an increasing amount of plastics. Once the recyclable materials are mixed up with the rest of our household waste, it becomes extremely difficult and very expensive to extract the useful resources.

So far, recycling at source – sorting out useful materials before they are thrown away – has depended almost entirely on the initiative of the householder. It's left up to you to sort out and deposit bottles in bottle banks or find a local collection scheme for newspapers and magazines. So at the moment recycling is a voluntary exercise undertaken by some conscientious householders on a piecemeal basis. Much more will be needed to achieve the government target of 50 per cent of recyclable materials actually being recycled by the year 2000.

One way to encourage more recycling is to offer financial incentives. For large-scale recycling to be financially feasible there need to be viable markets, both for the collected materials and for the products created from recycled waste.

Financial incentives from the government to industry could promote the use of more recycled materials.

Another way to promote recycling is to improve the facilities provided for recycling and information about them. This means more collection points such as bottle and can banks, and more experiments with recycling schemes which are sponsored by local authorities and businesses.

What can be recycled?

The five materials that make up most of our domestic waste – paper, food, glass, metals and plastics – all have potential for recycling.

Paper and card

Paper and card make up 25 to 30 per cent by weight of the average UK dustbin. Of all the paper and card used in the UK each year, around 30 per cent currently goes to be recycled. This might seem to be well on the way to the government target. But most of this recycled paper comes from commercial premises such as offices and shops and from industry. Only a tiny proportion comes from household domestic waste paper – mostly newspapers and magazines – that are collected through a network of voluntary schemes run by scouts, guides, local authorities and local environmental groups. There are also around 3000 paper 'igloos', similar to bottle banks, for old newspapers and magazines. Currently, over half the fibrous raw material used by UK paper mills is waste paper, and more is planned.

The recycling potential of paper and card is excellent. Most paper can be reused, although some paper products are difficult to recycle – paper drink cartons, for instance, are often made of a mixture of cardboard, plastic and in some cases metal foil, which makes recycling too complicated.

The problem with paper recycling – like most recycling

schemes – is that the price drops dramatically when there is a glut. Over the last year or so the price for collected newspapers and magazines has been low because of the success of collection drives. The price for good quality paper, collected mostly from offices, is still high. Demand for collected paper will increase as more paper mills able to take a high proportion of used paper are built and begin to operate.

Buying products made from recycled paper helps to encourage more paper recycling by creating demand. It is now possible to get lots of paper products made from recycled paper – some of the most widely available are writing paper, envelopes and note pads, computer listing paper, carrier bags and lavatory paper. And recycled paper is already used by the UK paper and board industry for a large proportion of paper and board packaging.

Pulp and paper production is a very energy-intensive process, but making paper from recycled stock takes around 40 per cent less energy than using virgin pulp. So recycling paper can save energy as well as cutting down the amount of waste that is buried in landfill sites. Using more recycled paper also cuts down on the amount of wood pulp (a primary raw material) and finished-paper products that need to be imported. And it reduces the amount of space that has to be set aside for growing softwoods, many of which damage or destroy natural habitats.

Organic waste

Organic waste makes up 20 to 25 per cent by weight of the average UK dustbin. Food and organic waste is biodegradable; it breaks down naturally. Over four million tonnes of naturally biodegradable kitchen waste are thrown away in rubbish bins every year. Nobody knows how much is recycled through garden compost heaps.

Compost heaps offer the best alternative to putting this type of waste into landfill sites. A compost heap is a small step towards recycling your waste yourself and, if everybody with a garden had a compost heap, there would be a significant impact on the amount of landfill space needed for rubbish and on the cost of waste collection and disposal. See Chapter 6 for how to set up compost systems in your garden.

If you don't have a garden, individual composting is obviously not practical. In some countries, for example parts of West Germany, the local waste disposal authority collects kitchen waste and makes compost on a large scale. They do this with paper and card, too. In this type of scheme every householder has two bins, one for organic waste and one for inorganic waste, which are emptied alternately. Another option is to provide municipal composting sites where organic waste could be deposited. The mature compost resulting from such sites could be used in parks or sold to gardeners.

Glass

Glass makes up eight to ten per cent by weight of the average UK dustbin. We use around six billion glass containers every year in the UK. In 1988 16 per cent of these were recycled, largely through bottle banks. The number of bottle banks has increased since then, and there are now over 4000. But our recycling record for glass containers is still well behind the majority of other European countries, as the Table on page 42 shows.

The recycling potential for glass containers of all types is excellent. For some glass containers there is an option that is even better than recycling – they can be reused directly. Milk bottles are reused on average 24 times. Some beer, cider and soft drinks bottles can be returned to the retailer for a refund. Returnable bottles are extremely cost-effective, saving on energy and raw materials and reducing the amount of waste created. Returnable bottles are unfortunately uncommon these days. They are unpopular with supermarkets because they take up a lot of valuable space, and, unless enough people return bottles, the system can

41

PERCENTAGE OF GLASS RECYCLED AFTER USE

The Netherlands   62%
Switzerland   47%
Austria   44%
Belgium   39%
Italy   38%
West Germany   37%
Denmark   32%
Turkey   27%
France   26%
Spain   22%
Portugal   14%
Great Britain   13%
Ireland   8%

1987 figures: The figures are likely to have increased by a few percentage points in most countries by now.

easily turn out to be uneconomic to run. EC initiatives have led to tough measures being taken against non-returnable drinks containers in some countries, such as West Germany, and may lead to an increase in the use of deposit schemes here.

It's not that the raw materials from which glass is made are scarce; there is enough silica sand, soda ash and limestone on Earth to make any number of bottles. But a lot of energy is needed to melt these raw materials down, far more than it takes if broken glass collected from bottle banks – called cullet – is used as one of the raw materials for making new glass. In fact, by using cullet there is typically an energy-saving equivalent to 136 litres (30 gallons) of oil for every tonne of cullet added to the furnace batch of raw materials. If the energy saved in the extraction of raw materials is added to the energy saved in melting down those raw materials, overall, using cullet produces a 25 per cent reduction in the energy needed to produce glass containers.

In the UK we still don't collect enough cullet from our own bottle banks to satisfy the needs of the glass-making industry. Using bottle banks can really help to provide a useful raw material for industry, but it's important that bottle banks are used sensibly and emptied regularly. Some bottle banks become surrounded by carrier bags full of glass containers that can't be deposited when the bank is full, and, if bottles are deposited and smashed late at night, the bottle bank site can become both an eyesore and a local nuisance.

To avoid any potential problems, the bottle bank code has been developed. The code is designed to help people use bottle banks properly:

1 Remove all tops and caps (otherwise the recycled glass can be contaminated and damage furnaces)
2 Put the bottles and jars in the banks in the correct colour compartments; green, brown and clear (if green cullet is mixed in with clear cullet at the glass furnace, it causes the clear glass to discolour)
3 Keep bottle banks tidy by disposing of any cardboard boxes or plastic bags in the litter bins provided
4 Don't put returnable bottles, like milk bottles or those on which a deposit has been paid, into bottle banks: these should go back to the milkman or shop in the normal way
5 Don't use the bottle banks at night, as the noise may disturb people living nearby.

Metals

Metals make up eight to ten per cent of the average UK dustbin. Most of the metal in domestic waste is food and drinks cans, along with aerosol cans, aluminium foil and discarded metal products like old saucepans.

It's estimated that each household in the UK uses over 500 cans every year. Cans are thinner and use less metal than

43

they did in the past, but this waste represents a huge consumption of energy and other resources.

Cans are made of either tin-plated steel or aluminium. You can tell which cans are aluminium because a magnet won't stick to them. The magnet attraction of steel cans has led to their successful recycling from domestic waste by some local authorities. Around 23 local authorities use huge electromagnets to extract steel cans from the general waste stream, after it has been collected. In this way eight to ten per cent of steel cans used in the UK each year are collected and sent for recycling – around 950 million cans in 1988.

Aluminium cans are less easy to identify for recycling as they are not magnetic, and fewer aluminium cans are recycled each year. However, there are voluntary collection schemes, and aluminium cans can be recycled through 'save-a-can' skips (which accept both steel and aluminium cans). Save-a-can skips operate in around 68 local authorities and are part of an industry-sponsored scheme that pays out money to charities according to how many cans are collected from the skips.

All metal production is a highly energy-intensive process, even though cans are thinner and less metal is used these days than in the past. Recycling can make great savings. Recycled aluminium saves 95 per cent of the energy that would be used to make the metal from bauxite, the raw material from which aluminium is derived. As a result, aluminium is valuable and commands a high price.

Labelling cans to show what they are made of would help recycling efforts. It would also help if more save-a-can skips were available. If you do have one near you, remember to rinse out cans and, if you can, squash them before you deposit them. Flattened cans take up less room and are therefore cheaper to transport.

Plastic

Plastic makes up seven to nine per cent by weight of the average UK dustbin. In industry, most plastic scrap and some products (such as film cassettes and car battery cases) are recycled, but virtually no domestic plastic is recycled. Plastic makes up around 35 per cent of the packaging we use in the UK – from cellophane (used as wrapping for cigarette packets, for instance) to PET (used for fizzy drinks bottles). Plastic packaging is more energy-efficient to produce and distribute than many other forms of packaging. It's also relatively energy-efficient to collect, because it's so light.

The potential of plastic recycling is quite good, although there are technical and logistical problems that have to be solved. It's particularly difficult to recycle domestic plastic because there are so many different types of plastic used for so many different products. Some products can be made of more than one type of plastic. The problem this causes is that if different types of plastic are melted down together, without being separated before recycling, they will react differently: some may burst into flame at a certain temperature, while others melt. This means that mixed plastic waste can generally be made only into low-grade plastic, which can be used for such things as fence posts.

Future recycling schemes, where plastics can be collected separately, may be helped by products being labelled with the type of plastic used. In 1989 the British Plastic Federation set up pilot schemes in Sheffield and Manchester to sort and process domestic waste plastic for recycling. These are the first plastic recycling schemes to be attempted in the UK. In the USA plastics recycling, through extensive labelling and collection schemes, is far more advanced.

Most plastic will not decompose at all once it is buried in landfill sites. Indeed, it is often the strength and durability of plastic products which contribute to their attraction as manufacturing materials. This strength and durability means that the potential for creating long-lasting reusable plastic products is very good, and may thus have an environmental benefit – but only if the plastics are reused and not thrown away. However, a lot of work has gone into creating biodegradable plastics that will decompose harmlessly in landfill sites. Although research has been successful at creating biodegradable plastics, many environmentalists argue that this is a blind alley, distracting attention from plastic recycling and encouraging disposability.

Where to recycle

The most fundamental benefit of recycling more of the materials that make up domestic waste is energy conservation. It is the saving in energy created by using recycled products to make new products (rather than using new raw materials) that most often makes recycling an economically viable as well as environmentally beneficial activity. It makes no sense to recycle domestic waste on a large scale if we use more energy collecting and transporting the materials than we save by reusing them.

For this reason it's important that recycling facilities are available locally and conveniently for everyone to use. If you need to use a car to make special trips to a bottle bank, rather than dropping off your bottles during the course of a normal shopping trip or school run, the energy used in petrol consumption may outweigh the energy saving represented by the collected materials.

To find out where your nearest recycling facilities are, contact your local authority. Local directories of recycling facilities have also been produced by Friends of the Earth and *The Daily Telegraph*. There are 51 of these directories covering the whole of the UK, available from W H Smith or Friends of the Earth (address on page 199). A national directory of recycling facilities has been published by Waste Watch, a watchdog organisation run by the National Council for Voluntary Organisations – see page 219 for the address.

Creating less waste

Perhaps the most important contribution we can make to the problem of what to do with the waste mountain is to create more opportunities for recycling. Householders can separate recyclable materials at home and deposit them at recycling facilities. Another option is for local waste collection authorities to introduce door-to-door collection of recyclable materials.

In Sheffield an experiment was launched in November 1989 to try to find out how practical it would be to collect from each house materials that are suitable for recycling rather than making the householder responsible for transporting them to recycling centres. Around 8000 householders were issued with blue boxes, in which they can store separately glass, cans, plastics, newspapers and magazines, and batteries. Every week these are collected by a special dustcart that has compartments to store the materials separately. These are then sent for recycling.

For most of us, recycling requires more effort, especially as facilities tend to vary enormously from one area to another. As a minimum, recycling paper and glass should be possible in all areas. And if you have a garden, just starting a compost heap could cut down the amount of waste you throw away by up to 25 per cent.

Packaging

Another way to cut down on waste is to be careful when you are shopping. It's estimated by the Industry Council for Packaging and the Environment (INCPEN) that almost six million tonnes of glass, plastics, paper, board and metals are used to produce packaging every year.

Packaging is essential in some cases: it keeps food fresh and uncontaminated, and helps reduce the amount of food that is wasted by preventing it from going off before it can be consumed. If all packaging were done away with, the choice and standard of products currently available in the

47

shops would be transformed. However, a lot of packaging exists simply to make a product look larger or more appealing – think of the amount of packaging involved with most boxes of chocolates. It's been estimated that an average £40 weekly family supermarket bill includes around £6 for packaging.

So what can you do if you want to lessen the impact of packaging on the environment? If you're looking for packaging which is more environmentally friendly, be prepared for disappointment. Just as there's no such thing as a truly environmentally friendly manufactured product, the same applies to the packaging that products are wrapped in. All packaging has some sort of impact on the environment, whether it's because of the amount of energy used in its production or because it's not easy to dispose of. The most 'environmentally friendly' packaging is no packaging at all!

Reusing packaging by buying products in refillable containers where possible cuts down on the total amount of packaging you use up. Although the availability of refillable packaging is fairly limited, you can find examples – The Body Shop, for instance, has for a number of years been running a scheme where you take back toiletry bottles to be refilled. Some retailers and leading detergent manufacturers are now selling detergents in refill packs.

Action points

- Avoid products that are over-packaged, especially where the materials used can't be recycled.
- Buy products that are packaged in refillable containers.
- Buy products like fruit and vegetables loose rather than pre-packed.
- Buy products packaged in materials that can be recycled fairly easily – and make sure you *do* take the waste to be recycled (but avoid making a special trip).
- Choose products labelled as being packaged in recycled materials.

- Return all returnable bottles, such as milk bottles.
- Cut down on the number of plastic bags you get at the supermarket checkout.
- Try to reuse plastic bags as much as you can, or, better still, use a proper shopping bag or basket.
- Buy in bulk to reduce the amount of packaging used. Just choosing the largest practicable size, rather than individually packaged cans of drinks or packets of crisps, say, helps cut down on waste.
- Try to choose products that are long-lasting and not designed to be thrown away when they break – pens and wrist watches, for instance. It's often better to pay a bit extra for a good quality product that will last a lifetime and can be mended than to buy a cheaper disposable version.

Disposing of rubbish that cannot be put in a bin

There are lots of things that you might want to get rid of that you don't want to or can't put in the dustbin, such as building rubble, furniture and garden rubbish. What can you do with them?

Waste disposal authorities are required by law to provide household waste sites (or dumps), where larger items of domestic waste, such as furniture or old cookers, can be disposed of free of charge. They aren't allowed to charge for domestic items, partly in order to discourage roadside dumping, but they can charge for commercial waste.

Household waste sites are ideal locations for recycling facilities, but the level of service varies enormously. The best offer bottle and can banks, paper igloos, engine oil collection tanks, facilities for reclaiming CFCs from used fridges, and collection points for saleable materials, such as rags. Other items of rubbish can be disposed of in different ways.

Building rubble

If you're carrying out major renovations and you can't take the rubble to a household waste site, you'll probably need to hire a skip. Check with your local council first, as some councils hire out skips more cheaply than commercial operators do. One or two may even offer you a skip free of charge if you can convince it that several residents in the street will be using it. If the skip is on the road (rather than in your front garden), you'll need a permit from the council. You'll also need to make sure it is well lit and not causing an obstruction.

Furniture

The local social services department may collect furniture that can be reused, or you could try charity shops. You could also try advertising it in a local newspaper; many offer free advertisements for cheap items (usually under £10). There is also a voluntary organisation, The Community Furniture Network (address on page 217), that co-ordinates information about local furniture collection and reuse schemes. For furniture that is really past it, if you don't have a car to take it to a household waste site, most councils run a home collection service, but many have waiting lists and will charge a small collection fee.

Clothes

Clothes are always in demand for jumble sales and charity shops. Even clothes that are long past reuse can be a valuable donation – Oxfam sends all unsold clothing from its charity shops for recycling. Rags can be used as a raw material for manufacturing a number of products, from carpet underlay to roofing felt. Old clothes also have a scrap value as rags or cloths.

Old cars

Old cars (and domestic appliances) may have a scrap value, however past it they are. You're more likely to get something for a car that is still drivable, but even non-runners are often collected free by scrap dealers or local councils.

Abandoned cars can be an eyesore and a nuisance, but it is often difficult to get them towed away by the council unless the owner can be located and agrees to let the car be scrapped. A statutory procedure which councils have to follow can mean long delays while enquiries are made to find the owner. If the owner can't be found, a notice is posted on the car (usually for a week or ten days) warning that the car will be towed away. If the owner responds and says that the car has not been abandoned there is very little the council can do, unless the vehicle is untaxed and on the road.

Hazardous domestic waste

You may be worried that some rubbish isn't suitable to be taken away with your normal collection service; batteries, garden chemicals and used engine oil have all caused concern that they might contaminate landfill sites and eventually lead to the pollution of water supplies. But it's often difficult to know what to do with such substances. Remember to be careful not to leave potentially dangerous substances you may be collecting for safe disposal, such as medicines, small batteries or chemicals, within reach of children.

Used engine oil

It's illegal to pour used engine oil – sump oil – down the drains. Apart from being a pretty nasty substance, contaminated with lead and other chemicals, it can clog up drains. Some household waste sites have safe oil collection facilities. If yours doesn't, ask for a waste oil tank to be provided. You could also try local garages; many provide an oil collection tank.

Used batteries

Old batteries contain heavy metals such as lead, cadmium and mercury which can escape into landfill sites as the batteries corrode. If they are incinerated, the smoke is a pollutant. You can purchase batteries with a lower heavy metal content, or, even better, use rechargeable ones. At present, recycling facilities for normal batteries are very limited, though more schemes (like the collection scheme in Sheffield) should be started in the near future. Recycling schemes for car batteries and hearing-aid batteries are more advanced. Some manufacturers of rechargeable batteries have introduced a postal recycling scheme.

Asbestos and other hazardous materials

For quantities of materials which are highly toxic, flammable or corrosive, such as asbestos, it's best to consult your local waste collection authority or environmental health office for advice on safe disposal. They may offer a free collection service. In London a free collection service is run by the London Waste Regulation Authority (address on page 218) which will collect flammable, toxic and corrosive substances and up to 50 kilogrammes (110 pounds) of asbestos from domestic users or charities.

Bonfires

If you have too much garden rubbish to compost and you can't take it to a household waste site, there is always a temptation to burn it in the back garden. However, burning rubbish can be both a nuisance for your neighbours and give off extremely unhealthy fumes. Burning plastics particularly should be avoided, as some can give off toxic fumes. If you really need to burn garden rubbish, try to keep it to an absolute minimum.

Batteries

More than 400 million batteries are used up and thrown away each year in the UK. Manufacturing these batteries uses up many times the energy that we get out of them when they are used. But more important from the environmental point of view is the damage that might eventually be done by the toxic metals contained in the batteries, mercury and cadmium.

Mercury

The UK battery industry has estimated that, in 1988, batteries thrown away in domestic waste contained 30 tonnes of mercury. Most of this is from alkaline-type batteries, such as Duracell and Ever Ready Gold Seal. (Ordinary zinc carbon batteries contain hardly any mercury, and hearing-aid batteries are usually collected for recycling.)

In this country waste goes into widely distributed landfill sites (see page 35). However, the problem is more serious in some Continental countries, and the EC has issued a draft Directive which will require the mercury content of alkaline batteries to be reduced to no more than 0.025 per cent of their weight by 1992. (Until 1989 most alkaline batteries contained up to 20 times this amount.)

The surge of interest in the environment led some manufacturers to attempt to market 'green batteries' containing no mercury. However, these were ordinary zinc carbon types (which normally contain hardly any mercury) rather than the higher-performance alkaline types, and buyers could be misled into thinking they were getting something special.

More usefully, some makers of alkaline batteries have been able to reduce mercury content well ahead of any EC deadline; they claim that there is no significant reduction in performance or resistance to leakage.

Cadmium

A way to get around the mercury problem is to use nickel-cadmium rechargeable batteries. These can save the resources and energy used in making batteries, because each can be recharged up to 1000 times. So, although they are quite expensive to buy (and you need a charger), you can save money in the long run – particularly in things needing a lot of power, such as larger radio/recorders.

But for the environment there is another snag: these batteries contain about 15 per cent cadmium. Cadmium is at least as toxic as mercury, and can cause damage when the batteries are eventually discarded. It is not possible to make this type of battery without cadmium, but – unlike with alkaline batteries – it's worthwhile recycling them. The EC Directive will encourage recycling, and two manufacturers – Memorex and Varta – have already announced recycling schemes. But they both depend on the user sending batteries to the manufacturer and paying the postage – so they are unlikely to be very effective.

Rechargeable nickel-cadmium batteries are widely used in 'cordless' appliances, from razors and drills to telephones and video camcorders. Usually, you can't change the batteries yourself, or, if you can, they come in a sealed plastic pack available only from the appliance manufacturer, so recycling is difficult. The draft EC Directive calls for built-in rechargeable batteries to be readily removable by the user. This should have the extra benefit of allowing you to avoid the often excessive charges made by service agents for replacing rechargeable batteries.

Other solutions

Mercury-type hearing-aid batteries have now been replaced by zinc-air batteries. These last longer and offer better value than mercury types, yet contain no toxic chemicals. However, once you strip off a seal to let the air in, you can't stop them discharging, so they are not suitable for most other uses. And some high power hearing-aids still need mercury batteries.

Batteries based on lithium compounds have already been around for some years for watches and calculators, and recently more and more cameras have been designed to use them – due to their long life and ability to charge up a flash unit very quickly. They are expensive but do not deteriorate when stored for up to ten years.

Lithium is not toxic but, because the pure metal reacts violently in the presence of water, there was some anxiety concerning its use. However, it has proved to be safe for consumer-type batteries.

Lithium batteries have not been available to replace ordinary batteries for most uses because each cell produced three volts instead of 1.5. One company, Ucar, claims to have solved this problem and plans to market 1.5-volt lithium batteries late in 1990.

Research is going on into rechargeable lithium-based batteries; these may become available in the next few years.

═5═

EAT YOUR GREENS

A hundred years ago Britain looked different from the way it looks today. In the countryside the fields were a variegated patchwork of colour, while cows and sheep dotted the landscape. Today, in many parts of the country, the hedgerows have disappeared, crops stretch for acres in uniform colour, and the calves aren't in the meadows for very long. What's happened to the traditional rural scene?

The answer is largely to be found in two world wars and a chemical revolution that determined both the need for, and the means to achieve, cheap food production on a massive scale. Today the government is paying farmers to stop farming the land: we don't need the food they can produce – in fact, we're only just seeing the reduction of huge surpluses throughout the European Community. A small minority of farmers are actively promoting a return to traditional farming methods – with no help from chemicals, artificial fertilisers, hormones or antibiotics. Branded by some as unrealistic idealists, organic farmers say their methods are better for the land, better for wildlife, better for our water supply, better for our animals and better for us.

Are they right? Can you help your environment and yourself at the same time by choosing organic food?

What is organic food?

Organic food is best described by the method used to produce it. Organic farming avoids the use of man-made fertilisers, pesticides, growth regulators and livestock feed additives. Instead, the system relies on crop rotations, animal and plant manures, some hand-weeding and biological pest control. There will be a large measure of self-sufficiency on an organic farm: using animal manures as crop fertiliser, for example.

In theory, any type of food – fruit, vegetables, cereal products, meat and dairy products – and drinks – wine, fruit juice, tea and coffee – could be produced organically. In practice, in the UK there is an emphasis on fruit and vegetables, with a small amount of dairy produce and livestock raised organically.

Eating organic for the environment

Pesticides and artificial fertilisers have been literally poured on to the land in the last few decades, resulting in abundant yields of cosmetically appealing fruit and vegetables. But pesticides are designed to kill. Although there's little evidence of harmful effects on humans, their sole purpose is to be toxic to insects, weeds and fungi which could interfere with the growth of the crop. At times, this action inevitably disturbs the natural balance of prey and predator. For example, herbicides that eradicate weeds thereby destroy the food source for insects, which in turn would have provided essential food for birds.

Likewise, intensive farming practice has tended to turn farmers towards unicropping – a dependence on a single crop pulled from the land year after year. The ethic of maximum production has been so promoted in the last 40 years that the shape of British agriculture has changed dramatically. It's difficult now for farmers to turn their backs on that ethic and accept the lower yields – and possibly lower income – of organic farming. Even within the government's own 'set aside' scheme, where it pays farmers to leave a piece of land unfarmed, there is nothing to stop the farmer working the remainder of his land even harder –

potentially compounding any environmental damage.

Even on small farms, organic farmers will have several crops at any one time, with every field going through a rotation that includes a fallow period when the soil can rest to restore its fertility. Constantly changing the crop is designed also to restore organic matter in the soil. This organic matter binds the soil together. When it's depleted, topsoil can be eroded and lost. If soil erosion continues year after year, whole fields and farms could become barren.

But more research is needed before the claims of organic farmers can be fully substantiated, and before organic farming can be proved to have a significant impact on the environment. Although organic farming has a far longer history than intensive agriculture, it is far less well studied. Only in the last ten years or so has organic production been under close and careful scrutiny. Questions of environmental impact are not easily resolved – and certainly results will not be available or theories proved in the short term.

Eating for health

Many major food retailers believe that the people who buy organic food products from them are either dedicated environmentalists or people who want 'to do their bit'. But from research conducted by *Which? way to Health*, it seems that, on the whole, consumers who buy organic food don't seem to be that bothered about the environment. In 1989 *Which? way to Health* surveyed over 1400 people who took part in choosing food for their household, and asked them if they had ever deliberately bought organic food. One in five said they had. When they were asked why, only one in six people said they bought organic food for environmental reasons, and only one in twenty said it was their main reason (see Table, right).

By far the most popular reasons given for choosing organic food were concern for health and belief that the food would be free of pesticides and chemicals. The environmen-

WHY PEOPLE SAY THEY BUY ORGANIC FOOD

Reason	% giving this as *main* reason	% giving this as *any* reason
Health	32	50
Free from pesticides	20	35
Taste	8	20
Curiosity	10	12
Environment	5	17
Animal welfare	<0.1	4
Economic	<0.1	2

tal benefits of reducing or eliminating the use of pesticides and scaling down agricultural production seem to be no more than an added bonus to most consumers.

Better for health?

If consumers buy organic food in the belief that it offers significant nutritional advantages over conventional produce then they could be under a misapprehension. There's actually very little evidence to support such a theory. Nutritional analyses have revealed only slight differences in the amount of vitamin C, protein, calcium, iron or potassium in organic and conventional fruit or vegetables.

Tests conducted in 1989 for *Gardening from Which?* magazine support the results of other research in this respect. Identical crops of carrot, lettuce, French bean and calabrese were grown on adjacent plots of land. One plot was farmed organically, and the other was tended using conventional garden chemicals and fertilisers. When samples of each crop were analysed for nutritional content, no differences could be detected between the French bean samples. With the other vegetables the intensively grown samples had slightly higher levels of sugar. Organic calabrese contained more vitamin C than its intensively grown counterparts.

The quantities of nutrients involved are always small, and it's doubtful whether they would make an appreciable difference to the average diet. And some nutrients are of course lost in cooking.

But it may be that the consumers who mentioned health as their reason for buying organically grown produce were thinking in terms of an absence of what are perceived as unhealthy substances – chemical residues.

Free of residues?

Forty-three per cent of foods recently tested by the Ministry of Agriculture, Fisheries and Food (MAFF) contained detectable pesticide residues. But the significance of that finding for health is largely unknown.

More than 400 different types of pesticides are approved by MAFF. But the legal limits set for residues of these pesticides in food (known as MRLS) are based on good agricultural practice and are not a measure of how much is safe to eat. Tests on salad vegetables by *Which?* in 1988 showed that residues were well within the legal limits. Equally, that doesn't automatically provide reassurance: long-term health effects can only, by their very nature, be detected over long periods of time, and in recent years some consumers have begun to challenge the safety of some pesticides or the wisdom of the maximum residue limits. There are three areas of concern:

- the 'cocktail' effect of consuming mixtures of different pesticide residues in the diet
- the accuracy or general applicability of the average diet used by MAFF to assess the likely total intake of pesticides. (MAFF set what is called an 'acceptable daily intake' of pesticides based on the expected maximum residues in particular products.)

- some older pesticides which haven't been reviewed since 1965.

While the vast majority of organic foods should be free of pesticide residues, this isn't necessarily the case for two reasons. First, although the sale of produce labelled 'organic' is usually preceded by a two-year conversion period during which farmers are not allowed to use any pesticides on their crops, some pesticides linger in the soil for longer periods and are sometimes detected in tests. In 1989 analyses by the French consumer magazine *Que choisir*, for example, found residues of a forbidden organophosphorous insecticide in a sample of organic carrots. Secondly, organic farmers can suffer the problem of pesticide 'spray drift' from neighbouring farms. In certain weather conditions some pesticide sprays have been found to drift over 50 miles. It is virtually impossible to guarantee that food is completely pesticide-free if it is grown anywhere in the industrial world.

As far as other residues are concerned, organic food cannot claim any great advantages. Testing for nitrate residues on comparable crops of organically and conventionally grown foods has frequently shown that organic crops fare only slightly better. This is because although nitrates are applied in artificial fertiliser, they also occur naturally in the soil. There are currently no nitrate residue limits set by MAFF.

Better tasting?

Results from taste tests comparing organic and conventional produce have been inconsistent – particularly for cooked food. Gourmets and lay people alike have had problems identifying organic food in blind tastings, and do not show a constant preference for it. Taste tests were conducted on the vegetables grown in the *Gardening from Which?* tests mentioned above. Freshly picked samples of each vegetable (calabrese, carrots, lettuce and French beans) were given to

61

18 amateur tasters who were asked to spot the difference between the organically grown and conventionally grown produce. Only with the French beans could the tasters detect any difference – but they couldn't agree whether it was the organic or inorganic beans that tasted sweeter.

Going organic

Counting the cost

If organic food cost the same as conventional food, there would be every reason for consumers to choose it: possible beneficial effects on the environment, possible avoidance of health risks and slight nutritional advantages. In other words, organic food won't do any harm and might do some good.

Unfortunately, you can expect to pay between 30 and 80 per cent more for most types of organic food, and occasionally as much as 300 per cent more. Price differences vary according to the season, whether the food is imported (about 70 per cent is), where you buy it and how processed it is. Processed foods like organic bread or biscuits tend to be less price-inflated in comparison with conventional foods, since the cost of raw materials is only a small part of the total price. Fresh organic fruit and vegetables are more expensive for three main reasons:

- yields are often 10 to 30 per cent lower
- there is more waste in the form of misshapen or diseased crops
- raw organic food has to be packaged to identify it and stop it being mixed with conventional produce sold loose.

Any savings organic farmers make by not using pesticides (a saving of perhaps £70 an acre for cereals) are swallowed up by labour costs and reduced yields.

Organic meat is expensive because in general the animals have to be kept for much longer before they attain the necessary weight. In intensive farming, for example, a four-pound chicken is ready for eating in 39 days, whereas an organically reared chicken is not ready for 80 days or more.

None of this augers well for a lowering of prices in the future. The major retailers claim they are not making any extra profit on organic foods. In fact, Gateway and Safeway have a deliberate policy of encouraging consumption by reducing the profits gained from organic food, compared with those from ordinary produce.

One consolation is that you could be getting more for your money, since organic produce contains up to 20 per cent extra dry matter by weight. This is because artificial fertilisers encourage lush growth, or more water retention, in conventional foodstuffs.

Getting hold of it

The 1989 survey already described discovered that the main reason why people have not bought organic food is that they've never seen it, or that it isn't sold where they shop. One in three people claimed this as their reason for not buying, while only one in four said the price put them off.

Until a few years ago organic food was available only in health food shops or direct from the farm. Today, however, there are over 800 supermarkets stocking it on a routine basis. Much of it has to be imported, since less than one per cent of Britain's farmers have turned to organic production methods. However, there are currently some limitations on what can be bought in these supermarkets. Although many have a wide range of fruit and vegetables, items like butter and eggs are available only from farms or by mail order. Only Safeway sells organic beef, while none of the super-markets sees a mass market for organic chickens.

Choosing organic food

Don't be put off by the appearance of organic food. Although most of the major supermarkets relax their normal

63

vigilance on cosmetic quality, they allow only a small leeway for minor blemishes and odd shapes and sizes. Over the last 40 years we've become used to a quite unnatural perfection in the appearance of our fruit and vegetables. Organic farmers generally don't produce enough to enable them to discard all the specimens that look less than perfect. But before you buy, check on freshness: organic food can look a bit tired sometimes since it often takes longer for it to reach the shops (packing facilities are rarely available on the farm).

Guaranteeing you get organic

One of the difficulties with organic food is that you cannot test to *prove* that it has been grown without the use of pesticides or that farmers have adhered to the strict rules of animal husbandry. The only way farmers and retailers can offer a guarantee is through a system of independent policing of their production methods – from growing and stocking through to packaging and processing.

The need for such vigilance has led to the proliferation of organic certifying bodies, both in the UK and overseas. Most of the major food retailers rely heavily on the Soil Association – the largest of the UK bodies – to do farm inspections for their fruit and vegetables. (Waitrose buys some fruit and vegetables from Organic Farmers & Growers as well.) Although the Soil Association aims to inspect annually farms which hold its symbol, it admits it has not always managed to do that in the past. Nevertheless, none of the supermarkets has ever had any reason to doubt the authenticity of its organic produce on the basis of its own pesticide residue testing.

Regulating and controlling the use of the word 'organic' is important: not only because consumers clearly expect it to mean 'pesticide-free', but also because conventional food sold fraudulently as organic could be offered at a lower price, ultimately cutting genuine organic food producers out of the market.

Until recently there was very little agreement on the detail of organic farming standards. If a case of suspected fraud had been taken to court by a Trading Standards Officer, the court would have had no concrete guidance by which to judge the validity of the claim.

That changed in May 1989 when the United Kingdom Register of Organic Food Standards – or UKROFS – was established. UKROFS is a board of people representing producers, retailers, academic agriculturalists and Civil Servants. For the first time this board brought together elements of the various UK schemes (Soil Association, Organic Farmers & Growers, Demeter and Biodyn) to formulate a set of standards which can be referred to as a baseline for defining what is organic. Organic Farmers & Growers produced much more detailed guidelines for their members in order for them to comply with the UKROFS standard. However, UKROFS falls short of the rigour applied by Demeter and the Soil Association – leaving these bodies anxious to preserve the identity and unique qualities of their own schemes.

Registering with UKROFS is entirely voluntary, and its existence does not wipe out the other organisations, which can continue to use their own symbols. Where an organisation such as the Soil Association does decide to register with UKROFS on behalf of its members, those members can use two symbols or one or none on their produce labels.

What does this mean for people who want to buy organic food and ensure they get the real thing? It's not acceptable that consumers have to keep tabs on all the different symbol schemes and the variations that are implied in them. The best solution for consumers would be a single symbol – with some force in law. Unless UKROFS insists that all organic-sector bodies register with it and use its symbol, this won't happen.

The intervention of the European Commission in this debate provides another complication. In 1989 the Commission produced a draft set of standards for organic production throughout the EC and a system of pan-European certification. The Commission did not propose to introduce a European symbol, although there would be a form of words that could be used by producers who were registered under the scheme. More important, registration is again entirely optional, and the standard proposed is far below those of the UK organic-sector bodies – including UKROFS. Pressure is now on the Commission to upgrade the standard.

In the meantime, when you're choosing organic food, look for one of the labels below.

Awarded to food from British farmers and foreign schemes conforming to strict organic standards. Man-made chemical fertilisers, pesticides and chemical foodstuffs are banned. Crops have to be rotated regularly to protect the soil. Farms are inspected annually to monitor crop protection and animal welfare. Several hundred UK farmers now have the Soil Association symbol.

Awarded by the Bio-Dynamic Agricultural Association (BDAA) to farmers meeting what are probably the most far-reaching organic standards in the UK. As well as banning chemical treatments and introducing rotations, farmers use special herbal and other biodynamic preparations to nourish the soil, and they aim for 90 per cent self-sufficiency in fertilisers and animal feedstuffs. They also take account of lunar and planetary influences in timing planting and harvesting. Only 20 UK farmers have the Demeter symbol, but it's common abroad, so you might see it on imports. All symbol-holding farms are inspected annually.

Biodyn

Also awarded by the BDAA (see above) to farmers taking a holistic approach – but it is used for food produced on farms which have been under conversion for two years and are awaiting the full Demeter status, which takes three to five years.

To win this symbol farmers must not use artificial fertilisers or chemical pesticides, and they are inspected annually both in the UK and abroad. Organic Farmers and Growers had to make some changes before registering with UKROFS. Unlike the Soil Association, OFG is involved in marketing organic produce.

Launched in May 1989, this is the latest symbol to appear. It will be awarded to farmers meeting very high organic standards – roughly equivalent to those of the Soil Association – and their farms will be inspected annually. It represents the official standard for all UK produce which claims to be organic, but, like all these schemes, it is voluntary. You might see 'produced to UKROFS standards' or a UKROFS registration number rather than the symbol itself. At the moment it doesn't cover organic wine production or imports.

Not quite organic

Free-range chickens and eggs do not qualify as organic. By law, free-range chickens have to have ten square metres of pecking ground and continuous daytime access to open-air runs. But, unlike organically reared chickens, there are no special restrictions on their feed or medication. These could include yolk colorants, in-feed medication and grain treated with pesticides. No animals reared organically can be given additives routinely (for example, antibiotics) to prevent disease or promote growth: ideally, all their feed should come from organic sources.

Conservation-grade products – cereals and meat, for example – likewise do not qualify as organic. Any meat labelled 'real', 'kinder', 'humanely reared' or 'additive-free' is not organic unless it is also certified by one of the organic-sector bodies. Meat and poultry described in these terms is produced in a controlled way – usually avoiding routine use of antibiotics or growth promoters in feed, and paying particular attention to animal welfare, but the feed does not have to come from organic sources. What's perhaps more important is that there's no legal definition of the terms used.

Finding out more

If you're interested in finding out more about organic farming you can do so by joining in. Working Weekends on Organic Farms (address on page 219) is an organisation that matches organic farms and volunteers. For a modest subscription fee it will arrange for you to go and work alongside an organic farmer for the weekend, with bed and full board provided (but it's *not* a holiday).

Action points

- Find out if your local supermarket stocks organic foods. Compare the range with that offered in other branches or different chains. If need be, write to the store manager (or ask to see him or her) to ask if the range can be extended.
- Write to the head office of the supermarket and ask for details about its programme for monitoring the use of pesticides in conventional produce. Ask whether it has plans to increase the range of organic foodstuffs available and to offer more price incentives to consumers.
- Look for items where the extra cost of buying organic is not prohibitive: mushrooms, bread, biscuits, for example.
- Find out if there is a local farm using organic techniques which can supply you with fruit or vegetables cheaper than supermarkets can.

═6═

HOW GREEN IS YOUR GARDEN?

Whatever shape or size your garden is, it is probably the piece of 'green space' you treasure most. Whether it is a fully fledged organic vegetable garden or simply a patio with pots of flowers, it is a very individual living area. The choices you make about what to grow and how to grow it will reflect your own particular lifestyle. Many of us like the idea of picking courgettes and strawberries fresh each day from a carefully tended plot free of pesticides. But how many of us have the time or the energy to put into it?

In gardening, as in other areas, the consumer has been bombarded with confusing products and information and urged to be 'green' no matter what the cost. There is a move away from the traditional philosophy of 'if it moves, spray it' and a rising interest in organic and wildlife gardening. The garden is your own patch of environment to look after, so it makes sense to treat it well. Even if you don't want to go totally organic, there are lots of things you can do with little effort to reduce your use of chemicals and to encourage birds and other useful predators.

Chemical changes

Many of our standard gardening practices, such as feeding the soil, tackling weeds and dealing with pests, involve the use of chemicals in one form or another. However, being concerned about the environment does not necessarily mean giving up chemicals altogether. Sensible use of chemicals in the garden does not present any danger, and the actual amount gardeners use is unlikely to contribute to overall environmental problems.

That said, there are many gardeners who have switched to organic methods with good results. It is worth asking yourself exactly what you need chemicals for, as there may be non-chemical solutions to your problems.

Safety first

A survey carried out by *Gardening from Which?* in December 1989 showed that lots of outdated or even banned chemicals are sitting in garden sheds up and down the country. Unopened chemicals have a shelf-life of about four years, but once opened they may deteriorate rapidly. It is illegal under the Food and Environment Protection Act 1986 to store or use any pesticide that is not approved by the government, so now's the time to check yours out.

Check out your shed

- The following chemicals are now banned. Make sure they are not contained in your products: ioxynil (in weed-killers), chlordane, DDT, aldrin, dieldrin (in pest killers) and mercury (in fungicide).
- Do not pour chemicals down the drain. Small amounts should be diluted with water and poured on to waste ground well away from ponds, streams and ditches. Banned pesticides or large amounts of any chemical should be dealt with by the local authority: contact the refuse or waste collection department, which will dispose of them.
- Make sure the chemicals you do keep are clearly labelled. Keep lids and tops tightly secured and the containers stored well out of reach of children and animals.

Pet care

Using chemicals on the garden won't just affect the insects or weeds they're intended for. Your pets, and of course any wildlife that is in your garden, will also come into contact with the chemicals. None of the insecticides or weedkillers on sale to gardeners are toxic enough to prove fatal to pets when they are used according to the instructions. However, they could make your pet feel ill or cause minor skin irritations, so keep pets indoors until the chemical has dried thoroughly. Always scatter slug pellets thinly. Take particular care near water, as fish are very susceptible to chemicals in the water.

Misleading the consumer

Never buy products that are badly labelled or not labelled at all – it is illegal to sell pesticides loose or in unmarked containers. Sometimes the manufacturers themselves are encouraging people to buy more chemicals than they need by packaging the same product in different ways. For example, ici Bug Gun for Fruit and Vegetables and ici Bug Gun for Roses and Flowers both contain the same pesticide, pyrethrum, at the same concentration.

There is also an increase in the number of green-sounding names that are used to describe new products. Terms like 'environmentally friendly', 'natural' and 'traditional' make it very difficult to decide which products are genuinely organic and which contain some chemicals. The consumer needs to be aware that these products may be no more 'friendly' to the environment than others and that each product should be judged on its actual ingredients. For more details on 'green' labelling, see Chapter 3.

Feeding the soil

Concern about the levels of nitrate in drinking water has called into question the use of fertilisers in agriculture. You can buy organic fertilisers which work slightly differently to man-made ones, even though they both give nitrate to the soil. It is unlikely that sprinkling a little man-made fertiliser on your garden will have any significant effect on overall nitrate levels in the surrounding environment, but if you want to go organic you will need to understand the difference between organic and non-organic fertilisers.

Plants need nitrogen for healthy leaves and shoots. They can absorb nitrogen from the soil only when it has been dissolved in water as a nitrate. Chemically, the nitrates produced by an organic and non-organic fertiliser are exactly the same. The difference is that most man-made fertilisers, like Growmore, release their nitrates immediately for the plant to take up. Organic fertilisers, like hoof and horn, generally release their nitrates more slowly.

Organic fertilisers are considerably more expensive than non-organic ones like Growmore. One way to keep costs down is to try to reduce the need to keep applying a fertiliser. This can be done by getting the soil into really good condition: if it contains plenty of organic matter it will retain moisture and nutrients.

Reduce the amount of fertiliser

- Select a manageable part of the garden and improve the soil by adding as much bulky organic material as you can get hold of. Garden compost is ideal (see page 83 for details of how to make it) or use stable manure, spent mushroom compost or used potting compost. This will improve soil structure and encourage the activities of soil organisms.
- On the rest of the garden, continue to use a quick-release fertiliser like Growmore when the plants can really use the nitrogen, for example on leafy vegetable crops.

- Use slow-release organic fertiliser like blood, fish and bone in long-term situations, for example when planting shrubs and trees.

Organic gardeners say that their fruit and vegetables taste better, although this is difficult to prove scientifically (see page 61). Any home-grown produce tastes better than shop-bought because it is fresher, but if you do garden organically you can be sure your produce is free of chemical residues.

Overall, it does not pay to take short cuts by simply substituting an organic fertiliser for a non-organic one – you could end up just paying a lot of money for no real improvement. Increasing the fertility of the soil will pay dividends, though, and eventually you could do without fertilisers altogether.

Combating pests

Not using chemicals in the garden is all very well, but what happens when the plants are invaded by armies of caterpillars and swarms of greenfly? In general, organic pesticides are not very effective in reducing severe infestations of, say, greenfly (aphids). However, there are some practical preventive measures you can take to keep pests at bay.

Preventing pest attacks

- Improve your plants' natural defences. Choose plants that are suited to the local soil and climate, and grow disease-resistant varieties such as 'Avoncrisp' lettuce and 'Maris Piper' potato.
- Dig over the ground in autumn. This exposes overwintering pests to birds and frost.
- Encourage natural predators (see page 78).

Decreasing the amount of pesticides you use should increase the insect and bird populations, allowing the garden to develop a natural balance of life. If you do have to use pesticides, only do so when the damage to plants becomes unacceptable – an odd slug-nibbled cabbage might be the price you have to pay for a chemical-free garden. It also makes sense to hold off spraying until there is an infestation rather than spraying 'just in case'. Where possible, target the particular pest you are trying to kill and choose a product that wipes out only that pest. ICI Rapid is a good example – it effectively wipes out greenfly but will not harm bees, ladybirds or lacewings.

A pest-by-pest guide

Aphids

Pick off infested leaves and shoots where practical. Derris and soft soap are organic pesticides approved by the Soil Association and the Henry Doubleday Research Institute. They work on contact, so spray them on to the pests direct. Both control aphids to some extent, but in *Gardening from Which?* tests none proved as effective as ICI Rapid (with pirimicarb) or Synchemical Py Garden Insect Killer (with pyrethrum plus an inorganic additive).

Caterpillars

Check plants weekly and pick off caterpillars and eggs. Derris and pyrethrum work as well as non-organic products, although they are more expensive because you need to keep reapplying them. A biological control spray is available, containing a bacterium, *Bacillus thuringiensis*, which attacks the caterpillars; this is sold as Bactospiene, Dipel and BT4000.

Slugs

Keep plots free of debris where slugs can hide. Protect young plants with collars 20 centimetres (4 inches) high cut from plastic drinks bottles. Provided there are no slugs trapped inside when you put them on, these should completely stop slugs from reaching the plants. Aluminium sulphate (Fertosan, for example) is less harmful to other animals than chemical slug killers but is expensive. If you do use metaldehyde pellets, scatter them thinly and only around vulnerable plants.

Codling moths
These can be a problem, tunnelling into the centre of apples and pears. A biological control is available which uses female attractants (pheromones) to lure male moths to their death, thus preventing breeding. Pheromone traps need to be hung near the tree from late May onwards.

Carrot fly
Carrot flies attack carrot, celery, parsley, parsnip and fennel. The best way of dealing with them is to prevent the flies from reaching the crop. As they fly at ground level, erect a barrier of plastic sheeting 45–75 centimetres (18–30 inches) high around the plot or cover the crop with spun fleece.

Cabbage root fly
Cabbage root fly attacks brassicas, especially young cabbages. Make collars from 15-centimetre (6-inch) squares of old carpet underlay. Cut a cross in the centre, extending one slit to the edge so you can slip them around the stems of the plants. These are less effective than chemicals but they do limit the damage.

Reducing pesticide use

Soil pesticides are persistent chemicals, and as well as killing off pests they may well kill beneficial insects.

- Concentrate on trouble-free plants – replace disease-prone varieties of roses which need continuous chemical support.
- Start seeds off indoors – young plants withstand pest attack better than seedlings.
- Use physical barrier methods where possible and pick off

the affected parts of the plant.

- Choose a pesticide that is specific to the particular pest and does not harm other wildlife.
- Avoid persistent chemicals, such as soil insecticides, which are bound to kill beneficial insects such as ground beetles and centipedes.

A thought for wildlife

Whenever you use pesticides in the garden, you should be aware of the possible consequences for wildlife. When spraying chemicals – even organic ones like derris – keep well away from ponds, ditches and streams, as even the minutest quantity can kill amphibians and fish. Avoid spraying open flowers during the day when bees are pollinating – early morning or evening is best. Slug pellets can be picked up by hedgehogs and ground-feeding birds, so use them sparingly. Using only artificial fertilisers will reduce the population of earthworms, partly because the chemicals make the soil more acid but mainly because earthworms need organic material to feed on. Boost your soil's fertility with compost and manure as well.

Tackling weeds

Using weedkillers on large areas of lawn or border will save you a lot of time and effort. But it is possible to maintain a weed-free garden without using chemicals:

- **Mulch** If you dig out perennial weeds and hoe off annual weeds at the beginning of spring, mulching can keep weeds at bay all year. Black polythene is the cheapest weed suppressor, but bark chippings, gravel or fast-growing ground-cover plants will look more attractive.
- **Spot-weed** On a small lawn you can treat weeds like dandelions individually either with a chemical spot weeder (*Gardening from Which?* recommends Elliot Pocket

77

Weeder or Spraydex Lawn Spot Weeder) or by digging them out with a sharp knife or hand-weeding tool.

- **Establish good mowing practice** Cut the grass when it reaches 4 centimetres (1½ inches) to encourage thicker grass and therefore fewer weeds. *Gardening from Which?* trials show that weeds are less of a problem when the grass is kept at a height of 2.5 centimetres (1 inch).

Encouraging wildlife

By simply cutting down on chemicals you will probably see an increase in wildlife. A garden full of birdsong, croaking frogs and shuffling hedgehogs is more than just a naturalist's haven – it is one of the best ways to win the war against pests. Natural predators, like insect-eating birds and mammals and amphibians, play an important role in keeping aphids, slugs and caterpillars at bay. No matter how small the garden, there is scope to encourage wildlife without turning the whole plot into a wilderness.

Hedgehogs

Hedgehogs are very common in gardens, although, being nocturnal, they are not often seen. They travel great distances searching for food, so several individuals may visit your garden in the course of a night. They eat mainly caterpillars, beetles and worms, but will also eat slugs – though probably not in great numbers.

The best way to encourage hedgehogs is to remove potential hazards. Ponds should have gently sloping sides or a slipway to prevent them falling in and drowning. Use slug pellets sparingly, if at all, and if possible collect poisoned slugs and put them in the dustbin. Make sure there is an overgrown corner or a heap of leaves for hibernation (but not a bonfire site). Putting out food will help hedgehogs to fatten up in the autumn ready for hibernation. Put out tinned pet food rather than bread and milk.

Frogs and toads

Frogs are more common in gardens, but toads are more useful as they eat greater quantities of slugs. Both will eat flies, grubs, beetles and woodlice, feeding mainly at night. Toads might take up residence in a dry-stone wall or a rock garden, and frogs will be happy in a garden with plenty of dense vegetation.

The most important factor is to provide a pond for breeding. Even a small one made from an old kitchen sink sunk into the ground will be adequate as a spawning site. If there is room, build a pond that is at least 60 centimetres (2 feet) deep in the centre (to prevent it freezing) but with a shallow margin of around 10 centimetres (4 inches). Young frogs and toads need plenty of cover as they start to leave the pond, so plant the edges with marginals like yellow flag iris. If you already have a steep-sided pond, make a ramp or a series of stepping stones to allow easy access.

If you have a pond but no frogs, contact your local county wildlife trust (contact the Royal Society for Nature Conservation – address on page 206). The wildlife trusts operate spawn swaps in spring, and many gardeners are happy to give away their surplus spawn. If you have fish in the ponds, take care; goldfish are very partial to tadpoles.

Blue tits

Blue tits are specialist caterpillar eaters. They also eat aphids, aphid eggs and leaf miners. As they raise 10 to 12 young at a time, they can use up a great deal of insect food.

As with most birds, the general rule is to put out nuts and fat in winter but to stop feeding them in summer so they'll seek out their own insects. You can also encourage them to take up residence by putting up a nest box. This should be on a tree or post 2–5 metres (7–16 feet) above ground and facing east, away from prevailing rain and direct sun. The box should have internal dimensions of 10 centimetres (4 inches) each way and have a hole in the front, 25 millimetres (1 inch) in diameter.

Song thrushes

Although they feed mainly on worms, song thrushes switch to snails and slugs in dry weather when the worms move deeper into the soil. Their natural habitat is woodland edge, so a garden with dense shrub borders is ideal. Provide water and food on the bird table in winter and avoid using worm killers on the lawn.

Swifts, swallows and house martins

These summer visitors feed on flying pests, scooping up aphids and midges to take back to their nestlings. Each bird accounts for several thousand insects every day.

There is little you can do to attract them, although a garden pond is useful as a source of insects, and they use mud for building nests under the eaves.

Ladybirds

Ladybirds feed exclusively on aphids. In spring they lay eggs near to where colonies of aphids are starting to build up. The more aphids there are, the more eggs they lay. As soon as the young hatch, they start feeding on the aphids.

The only way you can encourage ladybirds is to allow the aphids to flourish. Don't spray automatically, and if you do spray, use ICI Rapid, which won't kill ladybirds; other pesticides, even organic ones like pyrethrum and derris, will kill ladybirds.

Hoverflies, wasps and parasitic wasps

The larvae of these flying insects all feed on aphids or caterpillars. Adult hoverflies rely on nectar, so you can attract them by growing a range of nectar-producing flowers to ensure a healthy population. Mixing up flowers and vegetables can help by ensuring that they have access to aphids *and* nectar in close proximity.

Slow-worms

Slow-worms feed mainly on slugs. Although they look like small snakes (up to 30 centimetres (1 foot) long), slow-worms are in fact legless lizards and are totally harmless. They prefer sunny gardens, spending their time under flat stones. They will also hibernate in the compost heap from October to March.

Bats

Bats eat a large number of insects, including aphids and mosquitoes. If they are not already in your neighbourhood there's little you can do to encourage them. However, if you already have them roosting in your attic, remember they are protected by law and it is an offence to disturb them. If you are worried about them, contact the Nature Conservancy Council (address on page 218). They generally roost during the day and feed at dusk, and you can provide a bat box either attached to a tree or under the eaves of the house.

Creating a wildlife-friendly garden

Whatever creatures you decide to attract into the garden, there are some very positive changes you can make to create a more wildlife-friendly environment. This does not necessarily mean turning the whole plot over to wildflower meadow and native woodland. Even the most traditional patio, lawn and border plot can be adapted in a small way. With 150,000 miles of hedgerow lost since the Second World War and thousands of species under threat, gardens have become the last refuge for many plant and animal species. By encouraging a little more wildlife into the garden we can all do our bit for the conservation of those species.

Gardening for wildlife
- Build a pond. Incorporating a pond, no matter how small, into the garden is probably the single most useful

81

thing you can do for wildlife. It will provide a home for frogs, toads, newts and dragonflies, as well as a watering station for birds, hedgehogs and other mammals.

- Grow some native shrubs. Elder, hawthorn, wild rose and dogwood attract berry-eating birds. These are also species that are being lost from the countryside as hedges are removed.
- Start a compost heap. As well as providing a cheap source of organic compost, it will provide a home for slug-eating slow-worms.
- Leave an untidy patch. Set aside one area of the garden and leave it to get a bit untidy. Let nettles grow, and red admiral and small tortoiseshell butterflies may lay their eggs there. Piles of logs or leaves will provide cover for hibernating hedgehogs.
- Grow nectar-rich flowers. These do not have to be native wildflowers, but avoid highly bred F1 hybrids and double forms which tend to be low in nectar. Alyssum, sedum, marigolds, stocks, nicotiana and wallflowers are all high in nectar.
- Look after birds in winter. Keep a well-stocked bird table and, especially, a source of water. Grow berry-bearing shrubs and trees such as cotoneaster, rowan, holly and berberis.

Wildlife warnings

Don't be tempted to introduce any foreign species of wildlife into the garden as you are likely to do more harm than good. Exotic animals like the bullfrog and the American trapdoor water snail have been sold in garden centres for domestic ponds. It is legal to sell and buy these creatures, but it is illegal under the Wildlife and Countryside Act 1981 to release them or to let them escape into the wild. Bullfrogs in particular can get large and will eat native frogs, toads and newts. Be very wary of introducing any unknown species of wildlife, plant or animal, as it can compete with native species, which may die out.

Compost

One of the best ways of improving the structure and fertility of garden soil is to add plenty of compost. Making your own compost sounds like hard work and often only the keenest gardeners undertake it. In fact, it is relatively simple to make and certainly less time-consuming than you would think. Once established, the compost heap is an effective way of tidying up the garden, a good way in which to recycle some household waste and a cheap source of organic matter to use as a mulch and to condition the soil.

Choosing a compost bin

There is nothing mysterious about a compost bin – it is simply a container to hold the material neatly, keep it evenly moist and let in some air. It also helps if the bin can keep the compost warm and allow easy access to get the compost out.

If you do not have much material to compost, choose a small bin of 0.28–0.43 cubic metres (10–15 cubic feet). If you then suddenly have a big batch of material, it can be stored in black plastic sacks, tightly sealed until there is room in the bin. Gardeners with more space and more material to compost should buy a larger bin; these are less likely to dry out and tend to produce more uniform compost.

Bins are generally made of plastic, timber or wire mesh. What it is made of is less important than how the bin actually performs. Compost needs air for the bacteria and fungi to do their job, so make sure the sides have air holes. Too much air can dry out the compost and slow down the process, so a wire mesh bin or one with wooden slats with wide gaps needs to be lined with polythene. Don't forget to make air holes in the polythene.

Look for a bin with a lid. This will keep out heavy rain but will also ensure that the compost does not dry out completely. Make sure the lid is easy to put on and take off. Likewise, you will need to empty and refill the bin many times, so check that you can inspect the contents easily.

Buying guide

For reliable, trouble-free compost without any fiddling around, plastic bins are ideal. Choose one which is light and easy to lift. *Gardening from Which?* recommended the Essex Enricher and the Garotta bin. For a larger bin, the wire mesh Gap Display Compost Container is large and good value, though it needs lining and watering.

Some gardeners prefer the look of timber bins. Choose one with slats not too far apart and which can be removed easily. *Gardening from Which?* recommended The Larch Lap High Compost Bin. Timber bins are generally more expensive than either plastic or mesh bins

Another alternative is to make your own compost bin using recycled timber (see diagram, right).

Making good compost

In an ideal world, compost heaps would be made with a balanced mixture of waste all in one go. In reality most people use their compost bins to keep the garden tidy as well as to recycle waste, and add small amounts of material as and when it becomes available. Making compost this way is not bad practice, but the time taken to fill up the bin means that you will make roughly two batches of compost each year instead of a possible four.

- Site the bin in a sheltered site out of direct sunshine.
- Mix several sorts of plant waste together rather than putting on thick layers of one type. Mix vegetable peelings with grass clippings, rose prunings and leaves.
- Keep the material evenly moist. Don't, for example, pour on a washing-up bowl of potato peelings – drain off water first. Likewise, if adding lots of dry leaves you may have to sprinkle water over the compost before putting the lid back.
- Don't make a heap with more than 30 per cent of grass clippings – the result will be a wet, slimy, smelly mess.

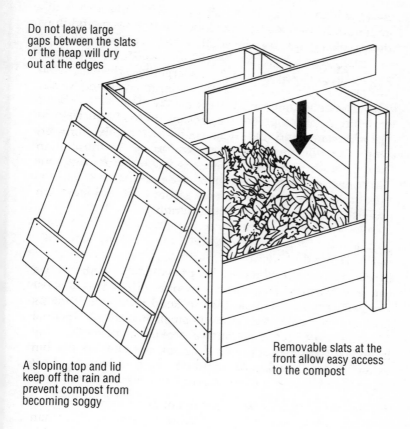

Do not leave large gaps between the slats or the heap will dry out at the edges

A sloping top and lid keep off the rain and prevent compost from becoming soggy

Removable slats at the front allow easy access to the compost

- Don't add diseased plants, weeds with seed heads or perennial weed roots.

Save time and effort

Gardening from Which? research shows that it is not essential to turn a compost heap. However, with small heaps where material tends not to get broken down at the top and sides, it will help to turn it occasionally. Likewise, an activator (like ammonium sulphate or dried blood) is not necessary unless there is a large quantity of woody material to break down.

Compost shredders

A compost shredder can convert woody material (like hedge trimmings, prunings and twigs) and fibrous material (like sprout stems and old herbaceous perennials) into fine chippings which will rot down more quickly on the compost heap. Apart from the expense involved in buying one, *Gardening from Which?* is not, at the present time, satisfied that all compost shredders are as safe as they could be.

The main drawback with shredders is that if used carelessly they could lead to accidental injury. One problem is the size of inlet, which in some cases is large enough to allow a hand to reach the cutting blades. Secondly, the protective funnel which would prevent this happening can be easily removed on some models. Thirdly, when the hopper is removed for cleaning the blades, the blades can still be operated – that is, some models have no device to disconnect the power when the hopper is removed.

If you would use a shredder only occasionally, such as after a major pruning, it might be worth hiring one. Make sure that it comes with full instructions, eye and ear protectors, and that the blades are properly guarded.

Substitutes for home-made compost

If you can't make enough garden compost, then you might want to find an equally 'organic' alternative, such as a manure or compost substitute. The problem with these bagged products is twofold. First, you need quite a lot of bulky material to make any difference to soil structure, so buying in bags can be expensive. One gardener described the small bags available as 'feeding strawberries to an elephant' – really too little of a good thing. Secondly, there is no legal definition of the term 'organic', and so-called 'organic compost' can vary widely to include fish meal, dried blood, seaweed and manure in differing concentrations. If you do buy organic composts, look out for the symbol of the Soil Association (see page 66), which approves only products that will be broken down by organisms in the soil.

Another way to boost your garden's organic matter is to find a cheap local source of farmyard manure. Contact a local farm or stables. Spent mushroom compost is also a good soil improver, and local mushroom farms will almost certainly sell it ready-bagged (but beware – it may contain pesticide residues and a lot of chalk).

Use less peat

Environmentalists are asking gardeners not to use peat. They fear that peat reserves are rapidly being depleted and that the process of peat-digging destroys the habitats of many rare plants and animals.

The peat we buy in garden centres comes from some of the most fragile environments of Ireland, Cumbria, Somerset and Yorkshire, although some is imported from Russia, Denmark and Finland. We could drastically reduce the amount of peat we use in our gardens, and this would significantly help the problem.

The alternatives to using peat are:

- for mulching: use bark chippings – they are more effective than peat
- for soil conditioning: use garden compost, if you have it, or well-rotted manure, spent hops or mushroom compost
- when planting trees and shrubs: use garden compost or reuse old potting compost in the planting hole
- reusing growing bags, which can be used for a second crop as long as the first crop was not diseased. The second crop should not be botanically related to the first, so tomatoes could be followed by cucumbers, herbs or runner beans but *not* peppers or aubergines. Growing bag compost can also be reused to grow annual flowers in

87

tubs and window-boxes, or to grow cuttings. It could even be used a third time to improve the soil when planting bulbs and perennials.

At present there is no proven alternative to peat for sowing and potting composts. The most promising development is the use of tree bark mixed with peat. You can use bark with a fairly clear conscience as it is a waste product of forestry. Other peat substitutes are making their way on to the market but have not yet been tested extensively.

Plants

It is difficult to tell if plants you buy have been taken from the wild or if they are a particularly endangered species. Certainly, you should not dig up plants from the countryside to put into your own garden – the ones growing naturally should be left where they are, and there are plenty of reputable nurseries selling wild flower seeds, bulbs and plantlets. With more exotic species, it can be difficult to tell where they have come from.

Bulbs from the wild

For years, bulb traders have dug up millions of bulbs from the wild and sold them to gardeners. If this practice continues, even common species like the snowdrop could become extinct. Turkey is the biggest bulb trader, exporting over 70 million bulbs a year.

Trade in rare plants is monitored by CITES, the Convention on International Trade in Endangered Species. The CITES agreement now covers cyclamen, snowdrops (*Galanthus*) and sternbergia. However, the regulations are difficult to enforce, and, in the absence of other sources of income, many countries continue to trade.

Positive action

Bulb traders themselves are taking action to ensure that bulbs taken from the wild are clearly labelled. They are also offering support to bulb cultivation projects in countries where alternative employment is needed.

Look out for bulbs currently under threat. These include wild species of *Galanthus* (snowdrop), sternbergia, cyclamen, *Eranthis* (winter aconite), *Leucojum* (snowflake), *Cardiocrinum giganteum* (giant lily), arisaema, narcissus and trillium.

Short of not buying these species bulbs at all, the best thing to do is to check that your supplier has a 'no wild-dug bulbs' policy. If he does not, ask him why not, or buy from another source.

Most bulbs sold in this country, certainly by the larger suppliers, are cultivated varieties of daffodils, tulips, hyacinths, crocus and so on. Whenever a bulb is labelled 'rare', it is worth checking that it has been cultivated. Apart from the fact that you will be helping to conserve the species, *Gardening from Which?* research shows that cultivated bulbs have a better chance of survival in British gardens.

The greenhouse

If you have a greenhouse in the garden or are considering buying one, you might be wondering how you can make environmentally sound decisions about how it is heated and how you grow plants inside it. Some of the problems faced by greenhouse gardeners are described below, with a look at how practical 'green' solutions might be.

Heating

Saving energy will probably be the most important concern for greenhouse users. *Gardening from Which?* research shows that after the initial cost of installing a power source, electricity is the cheapest form of heating. You need to aim for a temperature of 5°C (41°F) all winter to keep out frosts.

Saving energy

- If installing a new greenhouse, site it in a sheltered position, away from northerly winds but not shaded from winter sun. Bubble insulation saves 40 per cent on bills.
- Have an Economy 7 electricity meter installed – particularly if you can use other household appliances at night.
- Overwinter tender plants in a spare bedroom or frost-free outbuilding. That way you won't need to heat the greenhouse until March, when you start sowing seeds. If you buy young plants later in spring, you won't need to heat the greenhouse at all.

Watering the plants

In midsummer a well-stocked small greenhouse can use up to 45 litres (10 gallons) of water a day. As water metering charges are introduced in some parts of the country, this could be a problem. See also Chapter 8.

To avoid using tapwater:

- stop the greenhouse overheating by painting on an exterior shading paint like pbi Coolglass; keep the greenhouse well ventilated
- invest in a water butt to save rainwater
- grow plants in borders rather than in pots or growing bags. Borders need less watering (although there is a danger of soil-borne diseases)
- grow lots of leafy plants. The moisture given off by the plants' leaves will maintain humidity and keep temperatures down.

Organic options

Greenhouse gardening has its own set of problems, but it does allow you to grow a greater range of plants and vegetables than would be possible outdoors. For the organic gardener it is very tempting to try to run the greenhouse without using chemicals; although there are some difficulties, this is certainly not impossible.

Feeding greenhouse plants

One of the principles of organic gardening is to give plants enough nutrients to survive healthily without overdosing them with nitrates that can pollute the environment. As with garden soil, the best way of ensuring healthy crops is to build up a sound growing medium. If you don't want to use peat-based composts, use ordinary garden soil with lots of home-made compost or manure worked in.

Fighting pests and diseases

Caterpillars, aphids, red spider mites, mealy bugs and scale insects are all pests the greenhouse gardener has to contend with. Add to this the danger of fungal and viral diseases which spread rapidly in an enclosed environment, and it's no wonder that many gardeners reach for the chemical spray. However, there is a range of biological and organic alternatives to consider.

Biological controls work well in a confined space. Basically, a natural predator is introduced to prey on the pest. These are now available to deal with aphids, red spider mites, white fly and mealy bugs. Non-persistent chemicals like derris (made from the roots of derris plants), pyrethrum and soft soap are quick to biodegrade and do not build up in the soil. They can be used to treat aphids, white fly and scale insects. You could also try hanging up sticky traps, although these tend to catch beneficial insects as well as pests. Even organic pesticides can harm bees, hoverflies and ladybirds. Fungal diseases can be curbed using copper- or sulphur-based products. In a serious outbreak remove all the affected leaves, plants and compost from the greenhouse.

91

- Keep the greenhouse well ventilated. This will help reduce fungal diseases.
- Grow disease-resistant varieties of plants, such as 'Shirley' tomatoes and 'Carmen' cucumbers.
- Practise crop rotation. Don't plant botanically related plants in the same border each year.
- Wash down the greenhouse in winter with soft soap to kill pests.

Environmentally aware gardening

Gardeners are, in general, keen to look after their environment – if only because they spend a good deal of their time with their hands in the earth. However, there are some less obvious ways in which gardeners' decisions can influence environmental issues.

Wood work

Concern about the destruction of tropical rainforests might not seem to be relevant to British gardeners. But some expensive garden furniture is made from hardwoods such as teak or iroko which may have been felled from virgin rainforests. Manufacturers which use woods from properly managed plantations are listed in *The Good Wood Guide*, published by Friends of the Earth (see page 97).

If you dislike the idea of using wood preservative chemicals in the garden, choose oak, sweet chestnut or western red cedar, which are naturally resistant to rot. Even cheaper softwoods like pine and deal should last five years without preservative as long as they remain fairly dry. If the wood is going to be in contact with the ground, you need to soak the endgrain in preservative – brushing it on will not be effective. Soak the posts for at least 24 hours in creosote or an organic-based preservative: look for a government-approved Health and Safety Executive number or the Creosote Council logo on the label. Once dry, these chemicals are not harmful to plants, but as with any chemical you should follow sensible safety precautions.

Wood preservative safety check

- don't use it in an enclosed space as the fumes can be powerful
- wear gloves, goggles and old clothing
- leave treated wood for two to three days before use
- do not smoke, eat or drink while working with preservatives.

Preventing wood from rotting

- use naturally durable woods such as oak or western red cedar
- protect the tops of posts from rain with a cap of shaped wood
- place a piece of horizontal wood between fencing panels and the ground. It is much easier to replace this when it rots than the panels
- stand the legs of benches on paving stones or gravel rather than on soil or grass
- stand window-boxes and tubs on bricks to allow air to circulate underneath.

Gardeners are big consumers of wood products. Consider buying 'recycled' wood for your garden needs. This is usually wooden pallets which have been dismantled and had the nails removed. The wood is then resold to the public at approximately half the price of new wood. Contact Wood Chain in Birmingham on 021-550 8904. In other areas, ask your local environmental group.

Pollution in the garden

Unleaded petrol

If you run a petrol lawnmower, you could quite easily switch to using unleaded fuel. If you buy a new mower, you can start using unleaded straight away. Mowers which have been used for 50 hours or more on leaded fuel (two-star or four-star) should be decarbonised before changing to unleaded. Your local service agent can do this for you. Petrol-driven hedgetrimmers and chainsaws can also run on unleaded petrol, but check with the manufacturer first to find out if the engine requires any modification.

Pollution concern

Despite the growing number of vehicles using unleaded petrol, car fumes still represent a major pollution contributor. Some gardeners have expressed concern about the possible contamination of home-grown fruit and vegetables by lead. Lead tends to settle on soil and leaves but is unlikely to build up to toxic levels in the average garden.

If your vegetable plot is adjoining a very busy road, then it might be worth getting the soil tested. A thick hedge will provide it with some protection, and you could also use cloches or polythene sheets. Any lead deposited can be removed by peeling off the outer leaves and washing the vegetables using water with a drop of vinegar. Radishes and leafy crops like lettuce and herbs will absorb lead more readily, so avoid these if you think you have a lead problem.

Bonfires

The best way to clear up garden rubbish is not to burn it but to recycle it via the compost heap (see page 83). But for weeds and thick branches, having a bonfire might be the easiest solution. Save the material until it is quite dry and if possible use an incinerator. A hot fire which burns quickly gives off some carbon monoxide as well as other noxious gases. Do not burn plastic seed trays or flowerpots and never burn PVC, which can give off dioxin – one of the most potent carcinogens.

Recycling waste

In gardening, as in all other areas of our lives, making use of something that has already been used once rather than throwing it away helps to conserve the world's energy and resources. In the garden, old carpets can be used to suppress weeds, disposable plastic cartons can be used to raise seeds, and plastic bottles can be cut down to make slug protectors (see page 75). Look around your home and garden for 'rubbish' which could make a good plant container. The ideas are endless – from old car tyres to a ceramic lavatory cistern.

=== 7 ===

THE EMERALD FOREST

About half the world's tropical rainforests have been destroyed in this century – and the damage continues at an alarming rate. Tropical forests are vital to the environment. They help maintain the heat balance of the Earth's surface by absorbing carbon dioxide, a greenhouse gas responsible for global warming. They prevent erosion of the soil and prevent floods. At least 50 per cent, and possibly as much as 90 per cent, of plant and animal species grow or live in the tropical forests, and the forests are home to tribal peoples. As well as timber, many other products come from the forests, including fruit, nuts, pesticides, essential oils and spices, and drugs.

Why the forests are being destroyed

Tropical forests are being destroyed to provide land for farming, to yield firewood and timber, and so that mines and dams can be built. The importance of these factors varies from region to region. In many places logging is not the major cause for destruction; for example, in Latin America, clearing the forest for cattle ranching is very important. But even so, according to Friends of the Earth at least 12.5 million acres of tropical forest are destroyed by commercial logging each year. Usually only a few trees per hectare are cut for removal, but felling one large tree can pull down or damage others. Road-building, and removing the trees can also destroy the forest.

The effects on nature

Because loggers have concentrated on a small number of species of tree, some are threatened with extinction in some countries. Logging also affects other plants and wildlife in the area.

As well as causing damage directly, logging is also often the start of a chain that leads to the destruction of the forest. Loggers open up previously inaccessible areas by building roads and tracks, allowing colonists to follow and clear the land permanently. The reasons for this pattern include poverty and overpopulation.

What you can do

It may seem very difficult for individual consumers to have an impact on such fundamental problems. But you can use your purchasing power to encourage the timber industry to move towards using tropical wood that has been produced sustainably. If timber is managed sustainably, it should be taken from plantation forest or, if from virgin rainforest, the amount removed should be carefully controlled, and disturbance to the forest should be minimised.

In 1988 the UK imported 1.5 million cubic metres (53 million cubic feet) of tropical wood, plywood, veneer and logs. Among other things, this is used for furniture, doors, window-frames, boat-building, d-i-y and interior joinery. At present, less than 0.2 per cent of tropical moist forests are being managed sustainably for commercial products. Hardly any of the tropical timber on sale in the UK is from a sustainably managed source. Clearly, there's a long way to go.

If you want to avoid buying tropical wood that has not been managed sustainably, look for companies listed in *The Good Wood Guide*. This is published by Friends of the Earth and lists companies that have agreed not to sell tropical timbers unless obtained from sustainably managed sources.

They include furniture manufacturers and retailers, d-i-y and timber merchants and building suppliers. If you're not sure which types of wood come from tropical forests, you can check the list in the guide; they include mahogany, iroko, meranti, lauan and teak.

As well as looking for tropical wood from sustainably managed sources, you can also use alternative materials for your furniture or d-i-y needs – either wood from temperate forests or materials such as plastics, metal, concrete or stone. Tropical plywood can be replaced by a temperate alternative or other wood-based sheet materials from non-tropical sources.

You'll find these and other suggestions in *The Good Wood Guide* or in *The Good Wood Manual*, a more detailed publication (also available from Friends of the Earth). Woods from temperate regions can be substituted, particularly if they're to be used inside so that resistance to decay is not as important. For example, furniture can be made from timbers such as pine, oak, beech or cherry; doors can be made from pine. If you still hanker after the traditional deep colour of mahogany, temperate timbers can be stained to replace the reddish-brown of some tropical hardwood. Fitted kitchens can be made of chipboard surfaced with melamine or plastic laminate.

Alternatives to the woods used in garden furniture, such as teak and iroko, need to be durable enough to withstand the weather. You can use temperate wood that's been treated – but you need to be careful about chemicals that have been used, particularly for tables that will have food put directly on them. You could go for plastic or metal garden furniture instead.

If you really do want the mahogany-look for furniture and don't want to give up tropical hardwoods altogether, look for chipboard or MDF (medium-density fibreboard) covered with a veneer of tropical hardwood. But bear in mind that urea formaldehyde may be used to produce chipboard and MDF: there has been concern that exposure to this substance at high levels could be hazardous. Many furniture companies are now using low formaldehyde sheet materials and surface coatings.

Concerns for the future

Concern about the rainforests may mean that consumers must look again at wood. Although the colour and grain of tropical woods can be very attractive, consumers may come to appreciate the appearance of temperate woods more. Like fur coats, tropical hardwoods could become unpopular to be seen with.

Consumers will have to expect to pay more for wood: tropical wood is relatively cheap at the moment, considering that it may take many years to grow and is of very high quality. Managing forests sustainably and renewing resources cost money – consumers will have to be prepared to pay for this through higher prices for sustainably managed tropical timber. Some temperate woods, such as oak and cherry, cost more than some tropical woods at the moment.

The timber trade plans to publish its own timber guide, *The Real Wood Guide*, although the guide was not available as this book went to press.

What needs to be done

Action needs to be taken on an international scale to tackle the problem. The International Timber Trade Organisation (ITTO), which provides a system of consultation and co-operation between countries that produce and use tropical timber, is trying to do this. It helps to promote conservation and sustainable management of forests, although its powers are limited.

The ITTO is looking at whether an internationally agreed system for the labelling of sustainably produced tropical hardwoods is feasible. It's also been suggested that the ITTO

99

should adopt a Code of Conduct for the timber industry to ensure that tropical timber is produced sustainably. Such a Code might be difficult to monitor, though.

The European Parliament passed a motion in 1989 calling for the European Commission to control the import of tropical woods and to set up a fund to help timber-exporting countries to prepare forest management and conservation plans. The response of the timber industry to current concern about tropical forests has been to propose a tax or surcharge on imports of timber into the European Community. This would be used to raise funds for sustainable management of tropical forests. But this proposal has yet to be internationally accepted.

=8=

A DROP IN THE OCEAN

Water – a natural resource

Water is the most generous of our natural resources because it has the unusual property of continual regeneration. The clouds produce rain which, after falling on to land, either flows along streams into rivers and then the sea or is absorbed into the land and becomes part of an underground reservoir. This reservoir overflows, producing springs that add to the water in streams and rivers.

Once the water has reached the sea, evaporation returns the water to form clouds, leaving any impurities in the sea. Not only does the amount of water being recycled remain constant, but it is cleansed as well.

Why worry about water?

Such a natural recycling system ought to make a limitless quantity of water available to us. So why are environmentalists concerned about saving water?

The problem is caused by increasing demands on the water supply, particularly in cities and towns. Washing machines, garden watering, bath water and showers, swimming-pools – all put stress on the natural supply of water. As a result, in some parts of the UK we are now seriously interfering with the environment by leaving too little water in rivers or the ground.

But water is necessary to maintain the environment in the

condition in which we need it, both to retain our standard of living and the life around us on which we depend. If we need to take water from a river to meet our needs, we inevitably affect the environment directly around the river; the quantity of water further downstream will be reduced, changing the rate of flow of the water and altering the natural habitat of many plants and animals. And if the construction of a reservoir with a water-treatment plant is necessary, there's the question of how it fits in visibly with the surrounding land, and its effect on the wildlife which depended upon the river for its existence.

The apparent development of the greenhouse effect is likely to make the conservation of water resources much more urgent in some places. And perhaps less urgent in others. Seasonal variations in water availability could become more pronounced. A reduction in the amount of rainfall will lessen the amount of water available, and the shortage of water will make the demand for further resources much greater. Greatest increases in demand are expected to come from farm irrigation and garden watering.

The impact of reservoirs

Reservoirs have often been constructed without thought for their environmental impact, in areas which are important for their natural history and beauty. Thirlmere, in the Lake District, is such an example. The result of insensitive siting of reservoirs can be an expanse of water which has no sympathy for the surrounding area, and, in summer, an ugly scar revealed when the water-level in low. Conservationists are concerned about the siting of reservoirs because the resulting environment may not support the same variety of wildlife as before, and the landscape could deteriorate.

More recently, reservoirs have been constructed with much more consideration to their environmental effect. Chew Lake, near Bristol, is a good example of this, where the protection of wildlife, the enjoyment of the public and the appearance of the countryside have all been taken into account.

Taking water from under ground

At first, taking water from underground sources doesn't seem to be a problem. But if too much is taken, the effects can be serious in the long run. Life in a particular area depends on the water present in the soil – from crops to wild plants to animals and insects. Taking too much water can also create a vicious circle. For example, if a farmer's crop yield drops because the crop's roots aren't reaching enough water, the farmer may choose to irrigate the crop from above. This may save the crop, but in the long run it means that more water is taken from under ground, making it more likely that later crops will have to be irrigated too.

Abstraction from underground water also reduces the flows from springs. Many streams rely on this flow, and already abstraction has caused some to dry up completely. This has the knock-on effect of reducing the water entering rivers, thus slowing down their flow. Instead of springs feeding the river, there are places where the stream now disappears under ground because the water-level under ground has been lowered.

The drying up of the headwaters of rivers is becoming particularly noticeable in the chalklands of southern England. Near Luton, as long as 40 years ago, the headwaters of a number of rivers were found to be drying up due to increased abstraction of water from under ground. More recently, the urgent attempts by certain water authorities to overcome water shortages by increasing the number of licences to abstract has resulted in some abstractions being allowed without fully understanding the effect this will have on the environment. The newly formed National Rivers Authority now controls licensing for abstraction and should ensure that greater attention is paid to the environmental consequences of this practice. The effects on wildlife and farming of the recent increase in abstraction are already beginning to show, but will become more evident during the next decade.

The reduction in levels of underground water is happening widely where the ground is suitable for abstraction. The London Basin is an exception. It's a suitable site, and during the first half of the century the level of water in boreholes was falling rapidly. However, since then much less water has been abstracted by industry in London, and water-levels are increasing. Most of the water supply for London is now obtained from the rivers and the groundwaters outside the city.

What happens when a river slows down?

A river is a natural channel that has developed over millennia to remove excess water from the land. During this time, a river will have produced its own environment, on which a wide variety of wildlife has come to rely. The river is able to clean itself; in the winter wet weather generates fast river flows which scour out decaying water plants and deposits from the river-bed, leaving it cleansed for the coming growing season.

If abstraction takes place, or a reservoir is built, the scouring action of the winter rains is reduced. One of the results of this is to allow nutrients to accumulate in the river-bed deposits, and particularly in any lakes along the course of the river. When life starts to develop in spring, the change in the nutrients upsets the balance of the ecosystem, causing some organisms to accelerate their growth and others to die down. Plant life may become unbalanced, with excessive growths of brown or green algae, causing the water to become turbid. It is a noticeable trend in many rivers that the water is not as clear as it used to be. Although this may be caused by effluents, much of the turbidity appears to be natural growths encouraged by increased nutrients.

Who uses the water?

About half the water abstracted is used by industry (this doesn't include water taken for cooling nuclear reactors). But industrial use of water has decreased over recent years as water has become more expensive and as methods of production have been developed to use less water. On the other hand, the amount of water supplied for use in the home has increased by almost 20 per cent over the last 10 years and seems to be still rising.

Each person in an average household uses about 135 litres (30 gallons) of water a day. Around 43 litres (10 gallons) are used for flushing the lavatory, 23 litres (5 gallons) for baths or showers and 17 litres (4 gallons) for washing machines or dishwashers. The diagram below shows the use of water in the household.

There is no sign that the continuous rise in domestic water consumption will stop. This must mean that unless there are

Average household water use

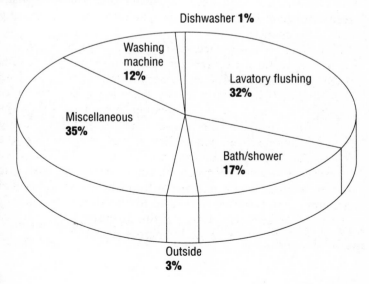

Dishwasher **1%**

Washing machine **12%**

Lavatory flushing **32%**

Miscellaneous **35%**

Bath/shower **17%**

Outside **3%**

major changes in the way we use water in the home, either new water resources will need to be developed or there are going to be restrictions on the way we use water.

The newly privatised water industry appears to have set its face against the imposition of restrictions such as hosepipe bans. The water industry will not be keen to encourage reduction in the use of water by individual households. Proposals for the widespread introduction of domestic metering will lead to an important source of revenue which will increase as water consumption increases. Metering, however, may have a different effect on the consumer. In order to reduce household bills it will be possible to lower the water bill by restricting the amount of water used in the home. A careful look at the domestic use of water by the consumer and what savings can be made will not only be good for the environment but also save the consumer money.

Where the savings can be made

The lavatory

The largest identified use of water is for flushing the lavatory. An important way of reducing this is now available with the introduction of the dual-flush cistern. Instead of nine litres (two gallons) of water being used with each flush, four litres (0.9 gallons) can be used when nine is unnecessary for hygienic purposes. This system could reduce water usage by about 13 litres (3 gallons) a day for each person – nearly 5000 litres (1100 gallons) a year, or a 10 per cent reduction in the water bill.

The bath

An average bath uses about 90 litres (20 gallons) of water, but a shower takes only about 30 litres (6½ gallons).

Someone who swapped their daily bath for a daily shower could save 420 litres (92 gallons) a week.

The washing machine

All washing machines use a lot of water – more than twice the amount you would use by hand washing. This does seem to suggest that there's some scope for washing machines to be designed to economise on water use.

There are big differences in water consumption for different models of washing machine. On a hot cotton wash (95°C/203°F), the best machines use between 71 and 85 litres (16 and 19 gallons) of water to wash a load of between 4 and 5 kilogrammes (9 and 11 pounds). The worst use more than 115 litres (25 gallons). On a synthetic wash (50°C/122°F) the best will use around 70 litres (15 gallons), the worst around 115 litres (25 gallons).

Tub-type top-loaders use a lot more water per wash than front-loaders or drum-type top-loaders. One model, in recent *Which?* tests, used around 160 litres (35 gallons) on a hot cotton wash.

You wouldn't normally think that a tumble drier uses water, but most of them do if they're washer-driers. Most simple tumble driers are air-vented – they blow the warm moist air out of the machine. Although there are a few air-vented washer-driers on the market, most are condenser: they use cold water from your house supply to cool the warm, moist air so that it condenses and can be drained away through the machine waste system, instead of steaming up the kitchen. Obviously, this adds to the total water consumption of the machine. The best condenser-drier will use around 25 litres (5½ gallons) of water per drying cycle, the worst around 70 litres (15 gallons). Also, most will dry only about half of their maximum wash load at one time, so to dry a full wash load these figures may almost double.

It might be possible for you to reduce the water you use in a washing machine by filling the machine with laundry to its capacity each time, or using the half-load button when necessary (if there is one).

For *Which?* recommendations of washing machines (and dishwashers – see below) that do reasonably well for both energy efficiency and water consumption, see Chapter 13.

The dishwasher

Like washing machines, different brands and models of dishwasher vary greatly in the amount of water they use – by up to 20 litres (4 gallons) on a normal wash programme. Some full-size dishwashers (taking 12 place-settings) use between 21 and 30 litres (5 and 6½ gallons) of water on a normal wash programme, whereas others use between 41 and 50 litres (9 and 11 gallons). Very few machines have economy programmes which save a significant amount of water.

As with washing machines, the most efficient way to run dishwashers to save both water and energy is to run them with a full load.

The garden

The use of water in the garden appears to be very small – only six gallons for each person each week. But this figure is an average over the whole country, over the whole year, and it hides a much more important factor in water use. Garden watering takes place over only a small part of the year in Britain, normally peaking in July. Its intensity is far greater in the drier south than the north and is obviously more intensive in towns that have a higher proportion of houses with gardens than flats.

Probably the most significant point is that watering occurs only at the time of the year when supplies are at their lowest. Water companies plan their water resources to ensure that there is sufficient water to cope with peak flows. In many parts of the country the peak flow is caused by garden watering, and consequently, if this can be reduced, the need to produce new reservoirs and underground abstractions could be reduced considerably. This is why most temporary restrictions on water use apply to the banning of hosepipes at the very time the gardener wants to use the water.

Hosepipe bans are not only to protect against a shortage of water. The intense demand for garden watering sometimes needs much greater volumes of water than can flow through existing water mains. Any plans for the water industry to reduce restrictions could mean a very expensive programme to install new mains just for this purpose. Once new mains were laid, the extra capacity could result in a sudden local increase in the demand for water supply to that area.

The only optimistic sign in reducing this water usage for the garden is that metering and the increasing cost of water might make people move back to collecting roof and surface water in underground tanks and in water butts.

Lost water

There appears to be a large loss of water after water leaves the water-treatment plant. Figures vary, but in one area it is said that 33 per cent of the water is unaccounted for and is presumed to be lost from leaky mains. To lose a third of all water does suggest serious defects in the water-supply system. Repairing leaks is time-consuming and difficult, and not something you as an individual can get involved in. But conserving water will reduce the need for new abstractions – if consumers can save water, then the water companies can also do so. You could try to find out from your local water company how much water is unaccounted for in this way, and raise it with the members of the regional rivers advisory committee of the National Rivers Authority.

Green detergents

There are now many brands of 'green' washing powders on

the market. These are supposed to be more environmentally friendly because they don't contain phosphates. Should you bother to change from your normal washing powder to one of the new brands?

What phosphates do to your wash

Phosphates act as water softeners and also help to suspend dirt in the water and generally improve the cleaning power of detergents. Phosphates aren't harmful to human health, but, after the detergent has been used, the phosphates remain in the dirty soapy water that is flushed down the drain.

Some manufacturers claim that using a phosphate-free powder will lead to a build-up of hard-water deposits on clothes in the long term, making them feel harsh to the touch and look grey.

What phosphates do to the environment

Phosphates are plant nutrients. In slow-moving water where they are able to accumulate, they can contribute to a process known as eutrophication, in which explosive growths of algae and other plants use up the available oxygen and eventually overwhelm the other life. There is now clear evidence of eutrophication damage in many UK waterways, especially where the water is slow-moving, as it is in the Norfolk Broads.

One way to deal with phosphates is to install phosphate-stripping equipment at sewage works. The equipment would largely remove both the 25 per cent of phosphates that come from detergents and an additional 25 per cent that is in sewage – from phosphates used in food and other products. But the stripping equipment would not deal with the 50 per cent of phosphate that goes straight into rivers and lakes from agricultural sources. Only a handful of UK sewage works currently have phosphate-stripping equipment.

Phosphate-free detergents

According to environmental experts, although using phosphate-free brands of detergent won't solve water pollution problems, it could help reduce the phosphate levels in some rivers. Using phosphate-free brands of detergent won't reverse the eutrophication process where it has already taken hold; nor will it stop further eutrophication, because phosphates can come from other sources. But using phosphate-free detergents is at least a start towards reducing the phosphate content of rivers and lakes until phosphate-stripping equipment can be installed at more sewage works.

The most common substitutes for phosphates in washing powders are silicate-based crystals called zeolites. They are not known to be harmful to the environment. Nitrilo-tri-acetic acid (NTA) is used in other countries as a phosphate substitute, but it has been linked to environmental problems because it can pick up heavy metals from sewers or river-beds, and could re-introduce them into the water supply.

Which? included phosphate-free detergents in tests of ordinary detergents to see how they fared. The tests cannot tell us whether the claim is justified that phosphate-free powders will eventually make clothes feel harsh and look grey. But they did show that phosphate-free powders, particularly those that contain bleach and enzymes, did well enough to suggest you should find them satisfactory for removing stains. So it's certainly worth trying them in place of your ordinary powder if your concern over the environment makes you want to change.

Some detergents which don't contain phosphates aren't labelled as phosphate-free on the packaging. Phosphate-free powders are generally more expensive than ordinary powders.

111

Biodegradability – a past problem

There used to be another problem with detergents and the environment, in that the surfactant contained in the detergents wasn't biodegradable. Since the mid-1970s most surfactants used in detergents have had to be at least 80 per cent biodegradable under European law.

9

ON YOUR BIKE

Transport, in all forms, is the source of several major pollutants in the UK. It produces 85 per cent of carbon monoxide and 45 per cent of nitrous oxides, both of which have a well-documented history of damage to the environment. And around 20 per cent of carbon dioxide, the largest single contributor to global warming, comes from transport.

Much of the blame can be pointed at the motor vehicle, which is responsible for:

- **carbon monoxide (CO)**, a colourless, odourless and tasteless gas. CO can cause death within minutes if inhaled within a confined space. CO converts to carbon dioxide, a greenhouse gas, when exposed to the oxygen in the atmosphere
- **carbon dioxide (CO_2)**, also a colourless, odourless and tasteless gas, not directly harmful to life but a principal cause of the greenhouse effect which is causing global warming. One litre of petrol produces about 2.2 kilogrammes (4.8 pounds) of CO_2
- **oxides of nitrogen (NOx)** Nitric oxide (NO) is another colourless and tasteless gas, which quickly converts into nitrogen dioxide (NO_2) in the presence of oxygen. NO_2 is a poisonous gas which destroys lung tissue. It is also, after it has combined with water vapour, a cause of acid rain
- **hydrocarbons (HC)** are basically the portions of fuel which haven't been properly burnt in the engine and subsequently are emitted from the exhaust. Some hydrocarbons may cause cancer and, when mixed with

113

oxides of nitrogen and sunlight, they can form oxidants which irritate the mucous membranes

- **lead (Pb)** is a poison which gradually accumulates in the body, eventually causing death when the lethal dose is reached. Lead is also claimed to retard children's mental development. In 1986, before unleaded petrol was available in the UK, nearly 3000 tonnes of lead were pumped out of car exhausts.

Atmospheric pollution is not the only problem caused by road vehicles, and cars are certainly not the only source of atmospheric pollution. Trucks, buses, and even trains, boats and planes add their share, not just to the air pollutants but also to noise, the huge energy costs of manufacture and the rapid use of fossil fuels. However, there are many steps that can be taken to mitigate the problems.

Lead in petrol

Strict British Standards exist for petrol sold in this country. Its volatility is defined, so that motorists can be sure that the engine starts easily and warms up and runs without problems. It must be clean, without any foreign bodies which may clog up fuel lines, pumps, the carburettor or fuel injection. It has to be free of water and have a sulphur content below a certain level. And the level of lead included has to be kept in check.

Lead has been added to petrol for decades. Its prime purpose has been to prevent pre-detonation of the petrol, or pinking. This occurs when the petrol in the engine ignites too soon in the cycle. It produces a rapid knocking noise but, more importantly, the engine runs badly, produces less power and is prone to overheating. The higher the octane rating of petrol – which in the UK is indicated by the two-, three- or four-star rating – the more resistant it is to pinking. Until recently the octane rating of petrol has been raised by the simple expedient of adding lead.

A second characteristic of lead is that it is a lubricant. This has been used to effect by engine designers, who have allowed for this property in the area of the valves, which in many cars have relied upon the lead to reduce wear. Without lead, the valves would wear rapidly, and frequent, expensive repairs would be needed.

Cutting back on lead – the problems

Lead from car exhausts has been dramatically reduced in recent years. Initially, the maximum allowable level of lead in petrol was cut back. Then, in the mid-1980s, petrol with no added lead was introduced to British service stations.

To overcome the potential problems caused by lead-free petrol, both the fuel industry and the car manufacturers have had to change their methods of production. Petrol now goes through a further refining process in order to reach the necessary octane levels without the addition of lead. Car engines are now designed to run on lower-octane petrol – whereas most cars on the market once needed four-star petrol, many can run on the equivalent of two- or three-star today. By using harder valves and valve seats within engines, the need for lead to provide lubrication in order to prevent rapid wear has now disappeared.

So the majority of new cars today have no need for leaded petrol, and by the end of 1992 every new car sold will have to run on unleaded fuel. But leaded petrol is still freely available, and will be for years to come so that owners of cars which can't use unleaded petrol won't be left out in the cold. But the advantages of using unleaded petrol are now so strong that everyone should use it, given the option.

Environmentally, the benefits are clear cut. But for many this hasn't been sufficient encouragement to get their car converted to run on unleaded petrol, so in the late 1980s the government introduced a two-tier tax system which currently makes unleaded petrol cheaper than leaded.

There has been some resistance to unleaded petrol, both by motorists and by those running fleets of company cars.

115

When unleaded petrol was first introduced, though many cars could use it there were often inconveniences. The ignition timing would need to be adjusted, a simple 15-minute task which was even offered free by a few makers, such as Vauxhall. But comparatively few drivers were won over – they either couldn't be bothered to get the job done, or were worried that they might be stuck with the adjusted car, unable to find a garage which sold unleaded. This last point is one of the several myths surrounding the subject – as long as a car isn't fitted with a catalytic converter, it can still be run just as satisfactorily on leaded petrol after adjustment for unleaded.

More understandable reticence can be put down to the requirement by some manufacturers that after the engine had been adjusted for unleaded fuel, drivers still had to make sure that, for example, one tank full in four was leaded. This, to make sure a bit of lead got into the engine to lubricate the valves, is clearly a nuisance, particularly as the result of not following the procedure correctly could mean an expensive engine repair bill.

This requirement has all but disappeared on the latest cars, while other disadvantages of unleaded petrol don't stand up to close scrutiny. It has been suggested that if you change over to unleaded petrol, the fuel consumption and performance will worsen. *Which?* carried out tests on four cars in 1988 to check this out. After the cars had completed their standard 8000-mile test programme on leaded petrol, they were driven for a further 2000 miles on unleaded petrol.

Two of the cars, a Ford Sierra and a Nissan Bluebird, needed some minor under-the-bonnet adjustments to allow the use of unleaded petrol. The remaining two, a Mazda 626 and a Renault 21, took unleaded straight away. The cars were then run through a series of tests for overall fuel consumption and performance.

Each of these tests was carried out in an identical manner to those which had originally been done with leaded petrol. In only one case was a minor difference in performance noted, but fuel consumptions remained the same (within the limits of experimental error). Perhaps as importantly, the four drivers who used the cars over the test period – the most experienced members of the car test unit – could detect no difference 'before and after'.

Another piece of hearsay suggested that unleaded petrol is a greater fire risk. The worry was that, in place of the lead, unleaded petrol might contain extra alcohol to raise its octane rating, and that this would change the flammability of the fuel. Normal fire extinguishers, designed to put out low-alcohol fuel blazes, might not be able to cope as well with this new fuel in the case of a car fire.

This worry has been discounted by numerous experts, including the London Fire Brigade. Tests have shown that even with petrol containing the maximum amount of alcohol allowed by EC law, there's no problem in extinguishing petrol blazes with existing firefighting equipment.

Using unleaded petrol

Any worries about the limited availability of unleaded petrol are now long gone. It is sold at virtually every petrol station, and in 1990 became the most popular grade of fuel. As well as the obvious environmental benefits, the financial savings gained by switching to unleaded can be significant if you use the car a lot.

Though new car buyers can generally use unleaded petrol without any difficulty, those with cars built before around 1988 may find it difficult to convert to the environmentally cleaner fuel. As mentioned earlier, adjusting a car to run on both fuels is a straightforward matter of altering the engine ignition timing, something quickly and easily done by any competent mechanic. It should cost less than £10, and may be free. The difficulty occurs when trying to decide whether or not this can safely be done to your car.

The way to find out, according to the majority of car manufacturers and importers, is to ask your franchised dealer, who should have the information at his fingertips.

But, as several *Which?* reports have shown, dealers are sometimes as confused as their customers, giving wrong or misleading advice, and there have been cases of the importer giving contradictory advice too. So if you are not entirely convinced by the information you are given by the dealer, write to the manufacturer's customer relations manager (addresses can be obtained from your dealer), outlining clearly the precise model details, including month and year of registration, and engine size.

Adding to the potential confusion has been the recent introduction of Super Unleaded petrol. This is designed for high-performance engines which need a higher-octane petrol than the 'standard' unleaded. The price is close to that of normal four-star despite the same tax advantage as given to standard unleaded because, the oil companies say, of additional manufacturing costs. If your car won't take ordinary unleaded it may run satisfactorily on Super Unleaded – check with the manufacturer, as above.

The RULE code

This rating system is designed to give unequivocal advice on the fuel a new car should use. Produced by the Department of Transport, every April and October, the RULE code should be available in car showrooms, or you can get one free from the Department of Transport, VCA Division (address on page 217).

The RULE code is:

R **REFER to dealer** before using unleaded petrol. This means that the engine can be adjusted or modified to use unleaded petrol, and/or may need to use leaded fuel every so often.

U **Must use only UNLEADED petrol** This invariably means that the car has a catalytic converter fitted.

L **Must use only LEADED petrol** These cars were designed to run on leaded petrol, and the use of unleaded petrol will damage the engine.

E **Can use EITHER leaded or unleaded petrol** These cars can be run on either fuel without the need for any adjustments, modifications or special precautions. Most new cars sold will be in this category.

Catalytic converter

Removing the lead from petrol – and hence from exhaust emissions – is only part of the battle against transport-derived pollution. Cutting down on three other main pollutants – carbon monoxide, oxides of nitrogen and hydrocarbons – can be effectively achieved only with a catalytic converter.

The drive against car pollution began in California in the late 1960s. There, pollution was so bad that a pall of smog would hang regularly over cities, obscuring the sun completely. Legislators tackled the problem by imposing tight limits on the amount of pollution allowed from a car exhaust, a policy which has been subsequently mirrored throughout the rest of the States, and in Japan, Australia and Switzerland. Now the European Community is catching up by tightening its own rules.

The solution in every case has been the catalytic converter. It's important to note, however, that nowhere is the catalytic converter specified as a compulsory fitment to a car. The catalyst is widely used simply because it is the only way currently known to meet the required standards.

What it does

The catalytic converter looks rather like the silencer box on the exhaust system, fitting into the exhaust pipe quite close to the engine. It converts the gases from the exhaust from CO, NOX, and HC to nitrogen (N_2), carbon dioxide (CO_2) and water (H_2O). Each of these occur naturally in the atmosphere and are less harmful than the exhaust gases. However, they do contribute to the greenhouse effect.

119

The structure of the catalytic converter comprises a stainless-steel shell with a honeycomb structure within. The honeycomb provides a large surface area for the exhaust gases to flow over, and this is increased even more by covering it with a wash coating which leaves an uneven surface. Though the converter is perhaps just 500 millimetres (20 inches) long by 150 millimetres (6 inches) in diameter, the surface area over which the gases have to pass is the equivalent of about three football pitches.

The important part of the catalytic converter is its final internal coating, mainly platinum, but also palladium and rhodium. These are the catalysts which cause the chemical reaction that converts the main pollutants into less harmful ones.

If this all sounds simple enough for us all to go out and get a catalyst bolted into our exhaust system, there are hidden complications. The first is that often there's not room. Many cars were designed when it was considered that there would be other ways of keeping pollution in check, and consequently do not have the space underneath to fit another box the size of a silencer without major structural – and consequently expensive – changes to the car floor.

Secondly, there's the matter of cost. When they were first offered on cars on sale in Britain, converters added as much as £800 to the price. Currently, the figure to the car buyer is more likely to be around £350, so they are still far from cheap, particularly when you consider that the basic unit costs the car maker around £50. However, even if you want to spend the money, it's not as straightforward as picking one up at the local exhaust centre. Only a few manufacturers (including Volvo and Rover) offer catalytic converters as a 'retro fitment' for older cars; your dealer should be able to help, though a call to the customer relations department is likely to give a more accurate response.

Varieties of catalytic converter

Today there are several types of converter on offer, some more efficient than others. The simplest is the two-way converter, which can reduce harmful emissions of HC and CO by around half, but won't have any effect on NO_2. They are in limited use today, generally on small-capacity engines. These engines have sometimes been tuned to be 'lean burn', which means that for any given quantity of petrol, more air than normal is mixed as it enters the engine. As long as the fuel–air ratio is very weak, there's a useful reduction in the amount of CO and NO_2 produced, though not necessarily in the HC emissions.

A three-way converter deals with all three pollutants by using a more expensive mix of noble metals as the catalyst. This is becoming the more popular choice in Europe, as it deals with around 90 per cent of harmful emissions.

Either type may be controlled or uncontrolled. Catalytic converters work most efficiently when the composition of exhaust gases produced by the engine remains constant. This occurs quite normally under certain conditions, but the mixture changes when, for example, you accelerate quickly. A controlled system has a small sensor inserted into the exhaust pipe just in front of the catalyst. This analyses, 1000 times a second, the constituents of the gases going into the catalyst; if they are less than ideal, the sensor will initiate changes in the ratio of air to fuel going to the engine so that the optimum balance is achieved. The cost of a controlled system is relatively high, not only due to the addition of the sensor, but also due to the fact that engines need fuel injection rather than the cheaper carburettor.

An uncontrolled converter doesn't have gas-measuring devices, so works at less than ideal efficiency for some of the time. Its great advantage, however, is that as long as there is room under the floor, it can be fitted to the majority of cars without any major or costly changes to the engine. In the next few years, with the tightening of regulations on emissions, it's likely that all new cars will have to be fitted with a controlled three-way converter.

121

Are converters a total solution?

A converter turns poisonous CO into 'harmless' CO_2 by the time it reaches the end of the exhaust. But carbon dioxide is a major contributor to the greenhouse effect. So how can the catalytic converter be justified in the overall scheme of things?

As far as CO_2 is concerned, the converter just speeds up a process which would otherwise occur normally. CO naturally converts into CO_2 in the presence of oxygen, so that – converter or not – we end up putting much the same quantities of CO_2 into the atmosphere. What the converter does do is ensure that the directly dangerous effects of the CO are not felt at ground-level.

So the only effective way to cut down on emissions of CO_2 is to use the car less, to drive at lower speeds and to accelerate more gently. And though car exhaust emissions play a significant part in global warming, they are certainly not the major contributor. Despite CO_2 accounting for a lot of the enhanced greenhouse effects, in the UK road transport accounts for only one-quarter of its emissions of CO_2.

A second argument concerns the actual effectiveness of converters on the road as opposed to that perceived in laboratory tests. The problem stems from the fact that the converter starts working only once its temperature has reached around 300°C (572°F) and doesn't reach its maximum effectiveness until perhaps 800 to 1000°C (1472 to 1832°F). Statistics from the Department of Transport show that the majority of car journeys are short ones, with the result that the converter rarely gets warm enough to operate effectively.

But even taking these points into account, a car fitted with a catalytic converter is certainly a significantly better environmental proposition than one without.

The drawbacks

While acknowledging the positive side to running a car fitted with a catalytic converter, the drawbacks should also be noted. The first is cost. While there is no legislation which forces the use of a catalytic converter before the end of 1992, most manufacturers are now fitting them to at least some cars in their range. Bearing in mind the horror stories in the mid-1980s regarding the likely cost (mostly put about by manufacturers who wanted to find alternative solutions to the catalytic converter), the additional bill doesn't seem excessive.

This may be because converters are being offered as 'loss leaders' by manufacturers keen to be seen as environmentally conscious. Volvo has offered customers the option of a 'free' catalytic converter on every model in its range since 1989. Porsche and Audi have fitted them as standard across their entire range of cars from the same year, with no apparent increase in the selling price of the car.

Other manufacturers may make a converter standard on an unusual or top-of-the-range model – Toyota was the first to do this in the UK with its four-wheel-drive turbo-charged Celica coupé. If you choose to specify a catalyst as a paid-for option on your new car, the current standard industry figure seems to be £350.

The second possible drawback is that an engine fitted with a controlled converter will suffer from a small power loss. This is because the converter is tuned to provide low emissions from unleaded petrol, rather than to produce the highest power. Whether you're likely to notice this difference is open to question. *Which?* tests of two 1.4-litre Fiat Tipos, one with a catalyst, one without, showed that subjectively drivers couldn't detect any difference between the cars. On the test track, the performance of the car with the catalyst was actually marginally better than the one without.

The same was true of fuel consumption. According to some estimates, fuel consumption will worsen by one to three per cent, which equates to £7 to £20 for the motorist

123

who does 10,000 miles a year. But since many cars with converters have more sophisticated engine management systems and fuel injection, the control over the use of petrol can be better than in the standard car, with the consequence that the fuel consumption might actually improve in the catalyst car. This was the case with the Fiats tested by *Which?*.

Catalyic converters and unleaded petrol

Cars fitted with converters have to run on unleaded petrol. Lead 'poisons' the catalyst, rapidly making it less effective at reducing the pollutants. To make sure you don't put the wrong fuel into a car fitted with a converter, the filler neck of these cars has a smaller diameter, so that only the nozzles of unleaded petrol pumps will fit in the orifice.

Unfortunately, the availability of narrow-necked pump nozzles has not kept pace with the introduction of converted cars, so, though unleaded petrol is freely available, it is often dispensed from wide-nozzle pumps which used to deal with two- or three-star leaded petrol. This situation should improve fairly rapidly, but in the meantime it is useful to carry around an adaptor to ensure that you can get unleaded petrol into your tank when you need to.

There have been some worries about the risk of contamination of catalysts through pumps which distribute both leaded and unleaded petrol through the same nozzle. But though the fuel apparently does come out of the same handle, there are in fact two separate nozzles inside; this ensures that there is no cross-contamination, short of the odd drip of petrol which may be left from the last user. Of course, there is always the slight chance that, despite the choice of fuel made at the pump, a fault may occur which means you get leaded when you want unleaded. To overcome this, it will soon be a requirement for unleaded petrol to be sold through dedicated pumps and nozzles.

EFFECT OF A CATALYTIC CONVERTER ON THE PERFORMANCE AND ECONOMY OF A FIAT TIPO

	With converter	Without converter
Car facts		
Engine power, bhp	70	72
Max. speed in 5th gear, mph	102.5	99.5
Acceleration		
Time to reach 60mph, sec	14.4	14.8
Time to cover ¼ mile, sec	19.8	19.9
30 to 50mph in 4th gear, sec	10.5	11.0
60 to 70mph in 4th gear, sec	6.4	6.9
30 to 50mph in 5th gear, sec	15.4	15.7
60 to 70mph in 5th gear, sec	9.2	11.4
Consumption		
Overall, mpg	37.7	34.7
Steady 56mph, mpg	48.0	47.8
Steady 70mph in 5th gear, mpg	39.2	37.8
Simulated city driving, mpg	19.0	16.8

Incentives to specify catalytic converters

The Swedes, Belgians, Dutch and West Germans are all offered tax incentives to run cars with catalytic converters. It may seem a little odd that the highly publicised benefits of converters are not enough on their own to stimulate a change in public attitude, but this is clearly the case even in a country as environmentally aware as Sweden.

Though repeatedly it has been hoped that, in his annual Budget, the Chancellor of the Exchequer would follow his European counterparts and offer tax incentives, this has so far failed to occur. That some compulsion or an incentive is required to boost the use of converters is illustrated by Volvo's experience: Volvo's offer of a catalyst on any of its new cars at no additional charge is accepted by only 50 per cent of its UK customers.

Looking after the converted car

To get the best from a car fitted with a converter there are a number of golden rules you should follow:

- Always use unleaded petrol. One tank-load of leaded petrol won't kill the catalyst stone dead, but it significantly reduces its efficiency.
- Start the engine according to the manufacturer's instructions.
- Do not race the engine immediately after a cold start. The converter isn't working until the exhaust has warmed up.
- Do not drive if the engine backfires, runs badly or lacks power. Pollution-levels increase when the engine is running badly.
- Do not tow-start the car. Lots of HC is put out through the exhaust unless the engine is running under its own power.
- Get the car properly serviced by a dealer who is equipped to cope with catalytic converters.

The family car

The use of unleaded petrol and catalysts is all very well, but there is no getting away from the fact that currently even the most environmentally friendly car still pollutes. So what can you do to maximise your contribution to a cleaner atmosphere?

1 Use the cleanest car possible

If you are buying a new car, it's likely that it can already use unleaded petrol without any problems. However, it's worth checking with your dealer before delivery and ask for any necessary adjustments to be made.

It is becoming increasingly likely that you'll also be able to find a suitable model fitted with a catalyst in your chosen range of cars. So pick a car with a converter, or specify one as an option. To get a catalyst model you may have to make a small compromise in the specification of your new car, but this problem is decreasing all the time.

Drivers running cars more than a couple of years old have less of a choice. Catalytic converters are available as a fitment on very few used cars (though Volvo is better than most in this area). So your main thrust should be to get your car adjusted to run on unleaded petrol. Vauxhall dealers will do this without charge (but only on a Vauxhall). Otherwise, your local franchised dealer will do the job, as will most garages and mobile tuning firms. Remember, though, that there are still a number of cars which cannot be adjusted to take unleaded fuel, so be prepared to be disappointed. Ask, too, when the adjustments are made, whether you need to follow any special conditions, such as using leaded petrol in one tank in every four.

2 Change your driving style

The faster you drive, the worse the pollution from your car. If you keep to moderate speeds, you'll not only help the environment but you'll also make a financial saving on petrol. And bear in mind that it isn't just high speeds that cause excess pollution – it's too much use of the accelerator at any time.

3 Change your use of the car

To many, this will translate into a change of lifestyle. The British seem to be following the American fashion of using the car wherever possible, rather than going by foot. But from an environmental point of view, cutting out the car for short journeys makes an enormous amount of sense. The pollution and fuel consumption are at their worst on short runs, particularly when the engine is started from cold.

Alternatives to the car

The healthy alternative

If you are really serious about cutting down on the environmental damage caused by transport, you should be considering alternatives to the car.

In order of lowest pollution, walking is best, followed by cycling, then using the train, bus and car. For many people walking and cycling are complementary. Those prepared to go by foot for a mile or so will be happy cycling two or three. Both demand a certain amount of energy on the part of the traveller, which in the majority of cases will be beneficial to their health, but there are disadvantages to be taken into account too.

Both groups, but especially the cyclist, are vulnerable to danger from motorised road users. Traffic fumes and noise are other drawbacks, though these will vary greatly according to precisely where you are moving about. On balance, however, there are an enormous number of opportunities to safely dispense with the car and do yourself some good at the same time.

Car versus rail versus bus

All of us need to travel by powered transport at one time or another, and the majority of travellers choose a car in preference to the alternative, for good reason. Cars take you from door to door, they leave at precisely the time you want them to, they are often more comfortable, and, with a certain amount of luck, you'll get to where you want to be fairly quickly.

It is for precisely these reasons that, by and large, public transport still finds less favour with the travelling public, except in certain circumstances – such as commuting into larger towns and cities. The late running of trains, cancellations, high fares, poor information, shabby accommodation and stations, and overcrowding are all criticisms that have been levelled at the railways. But while these criticisms all have some justification, there are many parts of the railway which do work well, providing fast, safe travel with time to work, read or relax – impossible in a car. And, of course, there are no problems about parking, which is a major difficulty in most city and town centres.

Buses too have their critics, with many of the same faults as the railways levelled at them. But both methods of public transport have major benefits in terms of the use of fuel and thus the amount of pollution. The Table below compares the use of petroleum (all oil-derived fuels, including petrol and diesel) by various modes of transport according to the number of litres used for every 100 passenger kilometres. Two figures are given, one for the vehicle with the number of passengers it typically carries, the second if the vehicle were used to its fully loaded potential.

PETROLEUM USED BY VARIOUS MODES OF TRANSPORT:
LITRES PER 100 PASSENGER KILOMETRES

	Typical load	Potential fully loaded
Express coach	0.9	0.7
Commuting bus	1.4	0.5
High-speed train	2.0	1.0
Moped	2.1	2.1
Minibus	2.2	1.2
Off-peak bus	2.8	0.5
Off-peak car	4.2	2.4
Motor-cycle	5.0	3.2
Train	5.4	1.2
Aircraft	9.0	5.8
Commuting car	9.2	3.0
Taxi	12.2	3.1

Alternative fuels

It would be nice to think that, hidden around the corner, there's a convenient alternative to petrol. This would still allow the convenience of personal transport while doing away with the problems of emissions. But the petrol engine has 100 years of concentrated development behind it, and it seems likely that it will be some time before a serious alternative is found.

Currently, the most obvious is the diesel engine, which is comparatively popular in mainland Europe and gradually gaining ground in the UK. Then, there's a very real possibility that an electric-powered car will soon become available (see page 131) – not a freak device but a development of the ordinary family car. As for other fuels, car manufacturers are known to be studying the possibilities, but they seem further off.

Diesel

In many ways a diesel engine is environmentally preferable to a petrol one. Diesel engines are inherently 'lean burn' as they operate at much larger ratios of air to fuel than conventional petrol engines. As a consequence, combustion is more complete, and lower levels of harmful emissions are produced. A modern diesel-engined car will emit far less CO, HC and NOX than a comparable petrol-engined car fitted with a regulated three-way catalyst. And, as a diesel uses less fuel, it produces significantly lower levels of carbon dioxide.

But there is a problem. Diesel engines emit particulates, basically sulphur, which contribute to acid rain, and soot. Some studies have linked diesel particulates to cancer, though others point to the fact that the concentrations, even in busy cities, are so low that the risk is unproven.

Diesel engine development is obviously aimed at overcoming these difficulties. Volkswagen is the first, and currently only, manufacturer to sell a diesel car which addresses some of these problems. Its Umwelt diesel, introduced in the Golf and Jetta models from 1990, is the first European diesel car to be fitted with a catalyst. This is responsible for a greater than 50 per cent reduction of the HC from a diesel exhaust. All other forms of harmful emissions are reduced too, aided by a turbocharger which ensures that the volume of air to fuel is kept high, and thus smoke-levels lower.

Electricity

Pollution created directly by electric cars is enticingly low – there are no emissions of CO, HC or NOX – so several electrically powered vehicles have been introduced over the years in an attempt to woo the public. Unfortunately, to date the only one to have any staying power in the market place is the milk float.

The reasons are down to the technology of the battery. To date, all electrically powered cars have had to rely upon the lead-acid battery, much the same as the one fitted to every car to provide power for starting and the electrical system. But the lead-acid battery has several major weaknesses.

First, in order to provide enough reserve capacity to ensure that an electrically powered car will cover a respectable range at a reasonable speed, you need a number of batteries. These, in turn, add an enormous amount of weight to the car and are more likely to take up more space than the engine/fuel tank of a petrol or diesel car.

Secondly, the batteries take hours to recharge, which is likely to be inconvenient, particularly when compared with 'recharging' your car with petrol, which takes just a few minutes.

Thirdly, batteries have only a finite life, so in time they will need replacement, which will not be cheap (nor particularly environmentally friendly).

Sir Clive Sinclair's attempt to sell a battery-powered, single-seater tricycle in the mid-1980s met with a famous failure in the UK. On a more practical note, Peugeot has announced an electrically powered version of its popular small hatchback which it intends to put on the French market in 1990. The electric Peugeot 205 has a claimed range of 75 miles at 45mph, a top speed of 62mph and accelerates from 0 to 30mph in 12 seconds. This compares with a range of 500 miles, a top speed of 97mph and a 0 to 30mph time of 4 seconds for the slowest petrol-engined 205. So even the latest electrical cars have a long way to go before they match the performance and convenience of their petrol equivalents. And that's before the matter of cost – Peugeot has said that there will be a 30 per cent price premium to pay, though partly offsetting this will be savings in bills for running and maintenance.

There is, however, a more serious problem which may hinder the acceptance of the electric car. UK power-stations are already responsible for the one-third of the carbon dioxide and NOx pushed into the atmosphere. Increasing the use of electricity to power cars will cut down on emissions on the roads but will have an adverse effect at the power-stations.

In the long term electric cars are likely to become a more common sight on our roads. But the real breakthrough depends on the development of an improved battery, and the generation of electricity in a more environmentally friendly manner.

Liquefied petroleum gas (LPG)

This has been fairly widely used as a fuel for vehicles, notably in Holland, Belgium and Japan. Since the gas is an explosive and an asphyxiant, the regulations for the storage and transportation of LPG are strict – in the car the LPG is stored in a steel cylinder, usually in the boot.

From an emission point of view, LPG offers minor improvements in HC and CO, while the release of NOx is little different. Fuel consumption is higher than for petrol.

Natural gas

Natural gas is not a viable fuel for cars in the foreseeable future. It has to be stored at 200 times atmospheric pressure, and then provides only a quarter of the range of the same volume of petrol. Emissions are similar to those of LPG, but the cost of adapting cars is more expensive owing to the expense of the high-pressure tanks.

Methanol

Methanol produces similar amounts of CO and HC to those produced by petrol, though the somewhat different forms of HC are considered to be less harmful. NOx are reduced, but against this is release of formaldehyde, which is an eye irritant. There are also safety risks.

Methanol's value as a fuel is limited by the need to carry twice as much of it as petrol for the same range.

Ethanol

This can be manufactured by fermenting biological compounds which contain sugar and starch. Ethanol has been popular as a fuel for cars in Brazil, where it was derived from sugar cane. But changes in the subsidies paid to farmers mean this is no longer cost-effective and the country has reverted to the use of petrol.

Emissions are much the same as those from a petrol engine, but acetaldehyde, which smells and irritates, is released in cold starting.

Hydrogen

Hydrogen is a fairly attractive prospect in terms of emissions, as CO and HC are eliminated from exhausts. NOx are increased, and there may be other emission products which cause problems. On a practical note, the difficulties of storage and safety are so immense that hydrogen doesn't look viable in the foreseeable future.

133

═10═

ENERGY FOR LIFE

In one sense, energy literally is for life – we need the energy in food to survive. Along with other animals, but unlike plants, we cannot directly use the sun's energy. But we use energy in many other ways, and life without it would be barely tolerable. In large parts of the world, where the main use of energy is for heating, life would be almost impossible without it.

Where the energy goes

Coincidentally, each year the UK spends about as much on energy as on food – a measure of its importance in our lives. In 1988 the country's energy bill was nearly £39,000 million.

There are different ways of looking at how a country uses energy. One approach is to consider sectors of the economy. Energy in the UK (and in most industrialised countries) is used in three main sectors in roughly equal amounts. These are:

- in homes – mainly for warming them, but also for hot water, cooking and for lighting and other uses of electricity
- in industry and commerce, producing goods and services
- in transport, moving people and goods around.

A second approach is to consider the final purposes for which energy is used. In the UK nearly two-thirds of all energy is used for heating, mainly warming our living and

**Spending on different types of energy by all UK consumers
(1988 figures)**

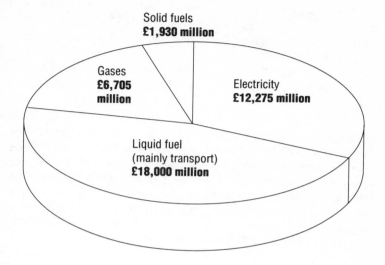

Solid fuels
£1,930 million

Gases
**£6,705
million**

Electricity
£12,275 million

Liquid fuel
(mainly transport)
£18,000 million

working spaces, but also for industrial processes. Just over a quarter is used in transport fuels, and about ten per cent is specialist use for lighting, electric motors, electronic equipment and so on, for which there are few convenient alternatives to electricity.

For detailed energy forecasting purposes, the two approaches mentioned are often combined, to give the purposes for which energy is used within sectors of the economy. The analysis soon gets very complicated.

Obviously, domestic consumers of energy have greatest control over energy use at home. The energy consequences of personal transport are also at consumers' command (these are discussed in more detail in Chapter 9). An important point, however, is that energy requirements occur in all parts of the economy – without adequate energy supplies the economy could not function.

135

How your use of energy affects the world

Modern energy systems are global in scale. European oil supplies may come not only from the North Sea but also from the Middle East, Nigeria or Venezuela; uranium for nuclear power-stations from Namibia; gas from Algeria or, increasingly, the Soviet Union; and low-sulphur coal from as far away as Australia. This has both good and bad consequences.

Formerly (and still today in developing countries), the constraints of distance forced communities to use local, often very poor quality sources of energy, such as fuel wood or low-grade coal. Now, the economies of industrialised countries demand high-grade energy and their consumers have a wide choice of supply in many cases.

On the other hand, any effects on the environment, because these occur a long way away, may be somewhat diminished in importance in energy consumers' eyes.

Environmental concerns about energy

Environmental concerns exist about energy use for two main reasons:

- the supply and use of all forms of energy involve the production of pollutants or the disturbance of land, which may directly affect health, disrupt wider eco-systems or limit enjoyment of the environment
- because some fossil energy sources are in restricted supply, we should use them in as sparing a way as possible to ensure that adequate amounts are available for future generations.

Generally speaking, many of the direct environmental effects of energy use on human health have been reduced to very low levels, at least when energy systems are operating normally. There are some exceptions – motor vehicle exhaust pollution in urban areas in some cases. Nowadays, attention is increasingly focused on indirect health effects, such as whether acid rain from fossil-fuel burning can increase levels of aluminium in drinking water, thereby causing some forms of degenerative brain disease. The effect of a man's activities on wider life forms are also receiving more attention.

One feature of both indirect health concerns and broader ecological concerns is that they are almost inevitably more complex to explain and have many uncertainties associated with them.

In the UK and other countries people are attaching greater value to the aesthetic quality of the environment. Wild, undisturbed areas are increasingly regarded as inviolable and not to be disrupted by man-made activities, of which energy-supply facilities are among the largest. Perhaps the most forceful example of this comes from Sweden, where, following a decision to phase out existing nuclear power-stations, the country faces a considerable electricity supply dilemma. Despite being a comparatively easy solution, the damming of four rivers in the north of the country to provide hydro-electricity has been ruled out as an option.

Although not strictly an environmental issue, a matter which is frequently discussed in comparing energy options is a direct risk to workforces and users associated with different energy options. Energy supply and uses are inherently risky, involving substances which are hot, under pressure, flammable, explosive, bulky or radioactive, and often need to be moved around. Everything from the risks in mining coal or extracting petroleum to the comparative dangers of gas explosions versus electrocution in the home have to be considered in this risk assessment.

The environmental effects of energy supply and use are very wide-ranging, from the local to the global. They also affect many different features of the environment.

Local effects

On the local scale the supply or use of energy have numerous consequences. Land may be affected in various ways. The degradation from the open-pit mining of coal or uranium, for example, can be severe. The land for wind-generator 'farms' may be less seriously affected, but its use will be restricted, and some people regard the visual effects as unacceptable.

Another local effect which fortunately occurs far less frequently nowadays in the UK is the smoke and sulphur dioxide smogs caused by burning coal in domestic hearths. As gas and low-sulphur heating oil have displaced coal, this pollution has declined.

Intermediate effects

Some air pollutants from energy use, although they may also have local impacts, create their main effects over widespread areas of land because to reduce local effects they are increasingly emitted from 'tall stacks'. Acid rain is the best known of these, affecting whole areas of continents, especially where the rocks and soils are already naturally acidic.

The dispersion of radioactivity from a serious accident at a nuclear power-station, as the Chernobyl incident demonstrated, can reach out to contaminate virtually an entire hemisphere, with weather conditions creating local 'hot spots' sometimes thousands of miles away from the site of the incident.

Global effects

Energy supply is mainly responsible for the possibility of global warming, since increases of carbon dioxide (the main greenhouse gas) in the atmosphere are largely the result of burning carbon-containing fossil fuels. Deforestation, which also contributes to the build-up of atmospheric carbon dioxide, is in part an energy problem – some deforestation occurs as growing populations in developing countries desperately seek fuel wood.

Some of the other gases which have a greenhouse effect come in part from energy sources. Methane, for example, as well as being emitted by rice fields and other wetland cultivation and by the digestive systems of grass-eating animals, also arises from the disturbance of coal seams in mining and (to a small extent) from leakages of natural gas during its distribution.

CFCS are powerful greenhouse gases. They also deplete the stratospheric ozone layer, another global environmental process, distinct from global warming. Responsibility for these gases cannot, however, be laid at the door of energy supply. Paradoxically, although their use is being reduced, CFCS were in the past used in foam energy conserving insulation, as found in the linings of refrigerators.

Most of the attention to energy–environment interactions tends to focus on air pollution issues. Energy use and supply have other environmental effects besides polluting the air, however.

Water pollution
Perhaps the best-known and most serious impact on water is the risk of contamination by crude oil spills, either leaking directly from offshore production platforms or, more commonly, as a result of accidents to oil tankers, of which the *Valdez* incident in Alaska provided a horrendous example.

Coal mining can pollute streams and lakes with the acid water often pumped out of pits. Nuclear power-stations, and the plants reprocessing their spent fuels, can also contaminate water if their fuel-handling or water-treatment systems develop faults.

Solid waste
Energy-producing systems can also create solid wastes which need careful management if they are not to cause

environmental problems. Coal produces by far the largest amounts of solid waste. Around one tonne in every three of material brought to the surface at some UK pits is solid waste for disposal, as the enormous spoil-banks in mining areas bear witness. In smaller quantities, a particularly unpleasant semi-liquid waste arises from preparing the coal for market and then, after burning, anything between five and twenty per cent of the raw coal remains as ash – not particularly hazardous, but needing disposal in large quantities at big coal-fired power-stations.

Nuclear power-stations produce far smaller amounts of solid waste – mainly the spent nuclear fuel rods taken from the reactor, but because of the highly radioactive nature of these, and some other solid wastes, they require very special handling. Most nations with nuclear power programmes are moving towards placing these wastes in carefully constructed underground 'repositories'; but some doubts remain over how effective this will be in ensuring that radioactivity will not return to the ecosystem over the very long time-periods involved.

Air pollution
Examples of several energy/air pollution issues have already been outlined. But others could be mentioned, especially photochemical smogs, closely related to the use of energy by motor vehicles. These first occurred in some American cities, such as Los Angeles and Denver, whose high levels of car use and local geographic circumstances made them particularly susceptible to the formation of 'brown haze'. Increasing levels of car use and periods of fair weather have made photochemical haze pollution a more common feature in other urban areas, including London. The pollution is caused by the action of light in a complex 'cocktail' of gases such as nitrogen oxides and unburnt fuel from car exhausts and fuel tanks.

Any energy chain has an impact on the environment

By no means every environmental effect of energy supply and use is listed above, but those that are should be enough to demonstrate that:

- all energy use has some impact on the environment
- energy use is implicated in many of the most pressing, pervasive environmental problems
- the scale of energy impacts on the environment ranges from the local to the global.

As soon as the energy–environment options are examined the problem is encountered of 'comparing apples and oranges' – which is the more environmentally undesirable, the visual intrusion of the 2000 or more wind generators that would need to be sited along the coast or the two or three coal mines and associated power-stations needed to supply the same amount of electricity?

An important principle in understanding the effects of energy use on the environment is to consider all impacts in the energy 'chain' – from the original source of the energy, through any processing or improvement, to the energy itself, its distribution to the final points of use and the effects of its final use – and also the energy and materials used at every point in the chain.

Environmental effects can arise at all stages in this energy chain. The energy chain approach also soon reconfirms that no form of energy is without its environmental impacts. For example, even if, realising a science fiction fantasy, we could supply all our energy needs by photo-voltaic assemblies mounted on the roofs of our houses and factories, there would still be the environmental effects of manufacturing the vast amounts of materials needed.

Reducing the environmental effects of energy by conservation must therefore be environmentally desirable. It is widely accepted that the prices paid for energy have not fully reflected the environmental cost arising from its use.

141

As higher environmental standards are demanded, the price of energy is likely to rise to pay for them. This will provide a strong incentive for conservation, but already in many cases we are failing, because of lack of information or institutional barriers, to save energy when it already makes economic sense to do so. We are therefore also losing out on the environmental benefits.

Your use of energy at home

Are there any environmental guidelines which you can follow in adjusting your own energy consumption? Nationally, the picture is that the domestic sector gets nearly 60 per cent of its energy needs from natural gas, 18 per cent each from electricity and solid fuels and 6 per cent from oil. These figures exclude energy used in personal transport. They are also, of course, average figures and do not reflect the pattern in individual households. The situation for individual households is determined by many factors, such as:

- whether piped supplies of gas are available
- the heating and hot water systems 'inherited' by people moving into a dwelling
- the up-front costs of switching from one form of energy-using system to another.

For some uses of electricity – lighting, running household appliances and electronic equipment – there is no feasible substitute.

The different uses to which energy is put in our homes are not directly reflected in the levels of pollution associated with them. This is because the different forms of energy used for different purposes have varying levels of environmental impact. For example, electricity-using appliances cause more greenhouse-effect gases to be released than are released when we heat our homes. Although warming our homes accounts for about 60 per cent of all domestic energy use, because this application is dominated by gas, which has a low production of carbon dioxide for a given amount of energy supplied, it produces less than half the total carbon dioxide emitted by the domestic sector. Electricity-using appliances and lighting in our homes, on the other hand, although accounting for about 10 per cent of total domestic energy use, are responsible for about 25 per cent of the carbon dioxide emissions attributable to the domestic sector because, in the UK, so much of our electricity comes from coal-fired power-stations and coal is a heavier producer of carbon dioxide.

The relationship will be much the same for sulphur dioxide, a major cause of acid rain, because once again gas contributes very little to sulphur dioxide emissions, while 70 per cent of the country's total comes from (mainly coal-fired) power-stations.

So, an environmentally sensitive household, while minimising its overall energy use, would meet its heating, hot water and cooking needs by gas (preferably heating by central heating, not direct gas fires) and would especially aim to keep electricity needs to a minimum.

Above all, however, it would avoid directly burning coal, even smokeless fuel, whose production merely transfers some pollution from point of use to point of manufacture and does not reduce carbon dioxide emissions. It is no accident that most of the areas in the UK where European Community air-quality standards for smoke and sulphur dioxide levels are infringed are areas where mining communities burn concessionary coal.

═══11═══

CENTRAL HEATING

Seventy-seven per cent of homes in Britain are heated by central heating or storage radiators. But do you know if your central heating is efficient, working properly and safe? When *Which?* took a heating expert around a variety of homes to inspect the central heating he found that many systems were inefficient – wasting fuel and money – much of the installation work was shoddy and that some systems weren't even safe.

Energy saving – good for you and the environment

Why is it important to make sure your heating system uses fuel efficiently? First, inefficient systems waste money. By reducing the amount of fuel you waste, you can make real savings in your fuel bills. Secondly, it's vital to use energy carefully to protect the environment.

The simplest way to reduce global warming and protect the environment is to reduce demand for all types of energy. Domestic heating makes up a significant part of overall energy consumption, so heating homes more efficiently could make a substantial contribution to conserving energy resources. Very often this can be accomplished quite simply too – by making sure your heating is well controlled and your home is properly insulated. A new system or boiler converts fuel to heat more efficiently than an older one.

How central heating works

The most common type of central heating is an open-vented, 'wet', fully pumped, smallbore, two-pipe system with an indirect hot water storage cylinder. But what on earth does this mean, how does it work, and how can you look after it properly to make it work to its maximum efficiency to save energy?

The boiler

The boiler is at the heart of the system. In all 'wet' central heating systems, water passes through a heat exchanger in the boiler and is heated up. This hot water is then circulated (usually by a pump) to the radiators and hot water circuits. The same water circulates continuously through the boiler and around the pipe system. This closed circulation is essential to reduce corrosion. Fresh water getting into the system could cause scale to build up, introduce air, causing corrosion, and dilute any corrosion-proofing chemicals.

Boilers can be free-standing, wall-mounted or back boilers (behind a gas fire or solid-fuel room heater). See page 161 for the latest developments in boiler design.

Maintenance

Most boilers need annual servicing by a professional. Older oil boilers should be serviced more often, because oil doesn't burn as cleanly as gas. A solid-fuel boiler should have its chimney swept once a year, or more often if soot tends to accumulate. Internal flueways and the pipe which connects the boiler to the main chimney will need sweeping every month when a solid-fuel boiler is in regular use. Electric boilers don't require regular maintenance.

The circulating pump

Modern systems are usually fully pumped; hot water is pumped around both the heating and hot water circuits.

This is a quicker and more efficient method than the older gravity systems, in which hot water from the boiler rises and circulates unaided simply because it is less dense than the cooler water returning to the boiler. Many older systems have pumped radiator heating with gravity hot water. Solid-fuel systems often still require a gravity circuit to act as a heat 'sink' – to stop the water in the boiler getting too hot.

Maintenance

If the pump is only on the radiator heating circuit (because the domestic hot water is gravity circulated), it won't be used for the summer months and may jam when you try to re-start it for the winter. To avoid this you could turn on the heating for a few minutes once a month during the summer. If the pump does jam, try tapping the pump body with a hammer. If this doesn't work it may be possible to free it using a screwdriver in the slot cut in the end of the driveshaft (turn the electricity supply off first). The driveshaft may be beneath a screw-on cap. If it still jams, the pump is probably blocked by scale and will need to be removed for cleaning or replacement. If the pump has been well installed there should be an isolating valve on either side, so that you can remove the pump without draining the system. If there are no isolating valves, fit them when you replace the pump. If you fit a new pump, make sure that the arrow showing the direction of flow points the same way as before.

Open safety vent pipes

There are usually two – one from the boiler, the other from the hot water cylinder – to act as a safety outlet if a fault causes the water to boil. It's possible to have a 'sealed' rather than an 'open-vented' heating system. In a sealed system there is no feed and expansion cistern and no vent pipes. The expansion capacity of the system is taken care of by a special expansion vessel and a safety valve. Most combination boilers are of the sealed type with the expansion vessel built inside the boiler casing.

Feed and expansion cistern

This is connected to the mains water supply and enables the heating system to be topped up with water when required. Very little fresh water is needed in a system with no leaks, so the main purpose of the cistern is to accept the normal expansion of water as the system heats up. It's also there to act as a place into which, in an emergency, any steam or very hot water can be discharged from the vent pipe. The feed and expansion cistern should have a separate overflow pipe.

Maintenance
Check for corrosion. If there is a lot of corrosion, empty the cistern with a bucket and clean it out. Check that the ball-valve is working properly – because it doesn't have to operate very often, it may have jammed closed. An empty feed cistern can lead to cold upstairs radiators, airlocks in the system and serious boiler damage. But the feed cistern shouldn't have much water in it – perhaps a couple of inches above the level of the feed outlet when the heating system is cold. The rest of the cistern is needed for the expansion capacity – so you don't lose the heating system water out of the overflow when it expands. Check that the overflow pipe isn't blocked and doesn't sag (so there's no possibility of trapped water freezing). If the cistern is in the loft, it should be insulated around the top and sides, but not underneath. It must have a lid. Any pipes in the loft should also be insulated.

Hot water storage cylinder

In an indirect storage cylinder the hot water for taps is heated indirectly by the hot water from the boiler passing through a coiled pipe inside the cylinder. Cylinders can also

147

be heated by an electric immersion heater as a back-up in case the boiler fails. With a direct hot water system, there is no coil in the cylinder and water is heated by an electric immersion heater, or typically a solid-fuel back boiler. In mains-fed systems cold water direct from the mains is heated and supplied 'instantaneously' (without storing it), for example by a gas multipoint heater, an electric heater or a combination boiler.

Maintenance
Make absolutely sure your hot water cylinder is properly insulated.

Cold water storage cistern

This is filled by water from the mains and feeds the cold taps (except the kitchen tap), wc and hot water cylinder. Recent changes in the Water By-laws have allowed purpose-designed and British Board of Agrément-approved hot water storage cylinders to be fed direct from the mains, doing away with the need for a cold water storage cistern. This is called an unvented system; such cylinders are not common in the UK but are widely used on the Continent.

Maintenance
Water is heavy; if the cistern is plastic, check that it is firmly supported across its whole base. It must have a lid to reduce contamination and evaporation, and a separate overflow. If it's in the loft it should also be insulated (though not under-neath). Check galvanised steel cisterns for corrosion.

Motorised valves

These are used to divert or stop the flow of water to the heating or hot water circuits when the cylinder or room thermostats register that the required temperature has been reached. They should be wired so that they send a signal back to the boiler to stop it firing when it's not needed. They should not be installed on or before the feed and expansion or vent pipe connections as this might block the expansion capacity of the system.

Two-pipe smallbore circulating system

This system has a flow pipe carrying hot water from the boiler and a return pipe taking it back. Each radiator is connected to both. This ensures that all radiators get hot water direct from the main flow pipe. In a single-pipe design – used in some older systems – the water flows from one radiator into the next, and only the last radiator in the circuit is directly connected to the return pipe.

Smallbore pipes are generally either 15 or 22 millimetres in diameter (most systems use both sizes in different places according to how much water they need to carry). Micro-bore systems have smaller pipes – between 6 and 12 millimetres in diameter – and each radiator has its own flow and return pipe connected to a central manifold.

Central heating controls

There's a wide range of central heating controls, but the point of all of them is very simple – to give you the required amount of heat or hot water when and where you need it. The more closely tailored to your needs your system is, the more efficient it's likely to be. To find out how well controlled your heating is answer the following questions.

Can you turn the hot water off when the heating is on?

In winter you may need the heating on all day, but only need to heat water for an hour or so in the morning and evening. But many older systems won't allow you to turn the hot water off when the heating is on. This is because they have a gravity hot water circuit with a pumped heating circuit controlled by a room thermostat. The room thermo-stat turns the pump for the heating circuit on and off. So with the programmer on and the room thermostat turned down you can have hot water without heating but not the other way around. To control heating and hot water on separate circuits you'll need to fit a cylinder thermostat, one or more motorised valves and possibly a better programmer.

Can you control the temperature of the hot water?

To do this you need a hot water cylinder thermostat, usually linked back to a motorised valve or to the boiler. This not only saves money by improving efficiency but is a good idea on safety grounds. It's easy to get scalded by very hot water from taps.

Is your boiler wired to prevent short cycling?

Short cycling means the boiler continually turns itself on and off to keep up the temperature of the water in the boiler and nearby pipes even though the house is hot enough and the cylinder is full of hot water. The boiler fires automatically because the water in the boiler itself has cooled below the boiler thermostat's level. This wastes a lot of fuel.

To check for short cycling: start with the whole system off and reasonably cold. Turn both cylinder and room thermostats right down. Then turn the programmer on. If the boiler fires it's not wired to prevent short cycling. This can be done by an electrician, who can correctly wire your heating controls (the motorised valves, room thermostat, cylinder thermostat, pump and programmer) back to the boiler control. More sophisticated boiler managers (controllers) that operate by sensing internal or even external temperatures are also available. See compensators and optimisers on pages 154–155.

Can you control the temperature in individual rooms?

Thermostatic radiator valves (TRVs) are designed to let you control automatically the heating in individual rooms. They make it easy to turn down the heating in (or avoid overheating) rooms that are gaining heat from another source – for example, kitchens and rooms that get a lot of sun. It's difficult to imagine any room where you'll always need the full output of the radiators, so TRVS are in principle a good idea throughout the house, although experts disagree about whether it's a good idea to fit them to every radiator – see page 154.

Timers and programmers

A good programmer should be accurate and flexible – so that it exactly matches your heating and hot water needs. Before choosing a programmer it's a good idea to write down a detailed account of how you use the heating and hot water for a typical week in the summer and winter — so you know what level of complexity of programming you need.

Older designs of programmers are often electro-mechanical – the type with tappets which you move around a clock face. A common problem with these is that the tappets are difficult to set precisely and won't move close enough together for a minimum time setting of less than an hour or so – though this isn't a major problem if other controls like a cylinder and room thermostat turn the boiler off anyway. Newer designs are mostly electronic – based on a microcomputer chip. These are easier to set accurately and can be more flexible than electro-mechanical programmers. Many can be programmed for each day of the week or have one programme for weekdays and one for weekends. Some, called two-channel models, allow separate timings for heating and hot water. This is particularly important because you may want the hot water set for the same times all year but different times for the heating depending on weather – you'd need two clock faces to do this with an electro-mechanical model. A good installer should help you choose a suitable programmer; ask him to show you two or three so you can practise setting them before you decide which to buy. Some typical features are:

Digital display
Either liquid crystal display (LCD) or light emitting diode

151

(LED). Make sure it's big enough for you to see clearly and placed at a convenient height and positioned out of direct sunlight.

Programming buttons
Often concealed behind a removable cover, these allow you to programme the unit's memory for the times you need. Most are set by a logical sequence (a bit like a video or digital watch). Follow the instructions carefully, but don't worry if you get it wrong – as a last resort you can always turn the power off or take the batteries out and start again.

Override
This allows you to change the settings for a short time – if you're unexpectedly at home, say – without losing your usual programmes. Programmers have various override facilities. The most common are 'boost' (this may be called 'extension' or 'extra hour') and 'advance' (often called 'change' or 'override'). Boost enables either heating or hot water to come on for a short time outside a timed period (this could be a one-, two- or three-hour boost). Advance reverses the settings currently operating when you press the button but automatically reverts to the programmed times at the next switching.

Battery back-up
This prevents you losing your programmes if there's a power cut. Some programmers operate on batteries alone.

Thermostats

Boiler thermostats
These are an integral feature of the boiler and are usually positioned on the boiler casing. Most have numbered settings rather than temperatures. It's best to keep the boiler thermostat setting high because boilers run more efficiently at high temperatures. But with gravity hot water which is

not controlled by a cylinder thermostat, the boiler thermostat will be the only way to adjust the temperature of the hot water.

Room thermostats

A room thermostat needs to be placed in a position that is not directly affected by any other source of heat gain or loss, such as sunlight, outside walls, draughts or radiators close by, so that it gets a reasonable typical temperature reading for the house. Experts disagree about whether you can have a room thermostat in the same room as a TRV. Some say that they will 'fight' for control and counteract each other; others believe that the room thermostat will receive a more typical temperature reading because the TRV is already moderating the heat requirement.

You'll need to experiment with a room thermostat to find a temperature setting which gives you a comfortable heating level in the main living-rooms. Don't be too worried about the actual temperature reading – it may not be precisely accurate (because it's in a draught) or relevant to the living-room temperature if it's in the hall.

Once you've found a comfortable setting, try to leave the thermostat to switch the heating on or off, rather than frequently altering the thermostat up or down as you feel hot or cold. Manually overriding the thermostat in this way will often give you too much or too little heat and is inefficient (a thermostat set at a comfortable level would have turned the heating off long before the living-room feels so hot and stuffy that you'd notice and turn it down manually).

Cylinder thermostats

These should be placed a third of the way up the cylinder and set at about 60°C (140°F). You can have a cylinder thermostat fitted independently (so it can be switched on in summer to improve the operation of a gravity hot water circuit), but to control the hot water all year round you'll usually need an electric thermostat linked back to a motorised valve and the boiler. Cylinder thermostats are

153

designed to have around a 6°C (11°F) temperature differential. This means that with a cylinder thermostat set at 60°C (140°F) it won't signal the boiler to fire until the temperature of the water has dropped to around 54°C (129°F) – low enough to be worth heating it up again.

Thermostatic radiator valves (TRVs)

TRVs work by reducing and finally shutting off the flow of hot water to the radiator when the required room temperature is reached. Those with horizontal heads or remote sensors are less likely to be affected by heat from the radiator or pipework than versions with vertical heads.

Don't lose the plastic pipe caps that are supplied with some TRVs. These allow you to close off the valve completely if you disconnect a radiator for decorating or maintenance. (Most TRVs open automatically if the temperature falls so low that there is a risk of freezing. So if you disconnect a TRV and don't fit the plastic cap you could end up with a flood after very cold weather.)

If you have TRVs on all the radiators and no other controls (like an overall room thermostat), there is no way to send a message to the boiler to prevent short cycling and no route for water to flow through if all the TRVs are closed. There are several ways to overcome these problems. You could have a bypass fitted (which could be a suitable radiator without a TRV) and a correctly wired room thermostat, or you could fit a flow sensor which samples the volume of water flowing through the system every few minutes and controls the pump and boiler. Alternatively, you could fit a boiler manager.

Boiler managers

Boiler managers are in many ways the most sophisticated central heating controls. While some boiler managers do little more than prevent short-cycling (which can usually be accomplished by some basic electrical work without the use of an expensive 'black box'), others, including compensators and optimisers, do much more than this.

Compensators directly sense both the outside temperature and the boiler temperature and use this information to compute the necessary firing time for the boiler, turning it and the pump on and off as required to maximise efficiency.

Optimisers allow you to set the actual times you require heating without having to make allowances for how long the house takes to heat up. An optimiser will automatically monitor your system over a week or so and learn from experience how long needs to be allowed for preheating. Then it will automatically turn the boiler on in time to have the house warm when you need it.

Installing a new system

If you're thinking of installing central heating, the first question to ask is: Do I really need it? In modern houses, particularly smaller ones that are well insulated, individual room heaters will be a lot cheaper and less disruptive to install and could be just as effective and economical to run. The Chart on page 158 shows running costs for different types of space heating – for the same amount of heat output (maintenance and standing charges are not included). Gas wall heaters, gas radiant/convector heaters or solid-fuel room heaters cost about the same for comparable heat output as most forms of central heating radiator systems.

Electric storage heaters

These are one of the main alternatives to full central heating. Storage heaters are usually sold as part of a 'Total Heating' package, comprising Economy 7 water heating, storage radiators and direct electric heaters like an electric fire or panel heater to provide top-up heat when required. Storage heaters work by storing heat from electricity supplied at the cheaper off-peak Economy 7 rate at night-time and then releasing it the following day.

Storage heaters tend to be more difficult to control than

155

conventional central heating because they give out a certain amount of heat all the time, whether you want it or not. You also have to decide in advance how much heat to store for the following day, which means anticipating the weather; if you get it wrong, this involves using expensive peak-rate electricity to top up the stored heat. The latest storage heaters can be fitted with thermostats and automatic heat input controls to help you allow for changes in weather conditions. Your local Electricity Board will be able to advise you on the latest controls and give you a quote for comparison with full central heating.

Storage heaters can't give the precise controllability of a gas or oil central heating system and tend to be most suited to smaller, well-insulated houses or flats which are occupied during the day.

Choosing a fuel

If you've decided on central heating, the next stage is to choose a fuel for the boiler. The Chart on page 158 shows the comparative running costs for one kilowatt hour (kwh) of space heating using different types of heaters and fuels. Apart from running costs you'll also need to consider:

- **capital costs** See page 157 for advice on obtaining quotes
- **storage space** Needed for LPG (propane), solid fuel or oil
- **availability** Mains gas isn't available to 13 per cent of homes in the UK
- **ease of installation** Storage heaters or individual room heaters often come out best here. There's less disruption from laying pipework, so the cost may be considerably less
- **how clean and convenient the fuel is** Some fuels will need more maintenance – for example, an unautomated solid-fuel boiler will need to be refilled and the ashes removed regularly
- **how controllable the fuel is** Gas, either mains or LPG

(propane), and oil tend to be more easily controllable than solid fuel or storage heaters – they give out more of the heat as and when you want it.

Gas is the most common fuel source for central heating – around 76 per cent of existing systems and 80 per cent of new installations are powered by gas. For most people, gas will be the best choice.

Finding an installer

A good installer is essential, and it pays to take some trouble to find one. You need someone to advise you on the latest equipment and give you an accurate and detailed quotation. A good installer will take the time and trouble to make sure the system really suits your needs. Personal recommendation or local knowledge are the best ways of selecting a shortlist.

Get three estimates for comparison and be wary of firms that offer only a standard package deal or concentrate on equipment they make. Include at least one independent installer on your shortlist. The organisations listed here can help you; most of the trade associations operate codes of fair trading or warranty schemes. For the addresses, see Useful Addresses, page 216. From March 1991 it will be compulsory for all gas installers to register with a new safety body, which will replace the voluntary CORGI scheme.

- **Confederation for the Registration of Gas Installers (CORGI)** By law, gas installers must be 'competent' to carry out any gas-related work. At present the best indication of 'competence' is that an installer is CORGI-registered. CORGI-registered installers undertake to provide work that complies with Gas Safety Regulations and British Standard Codes of Practice relating to gas safety. This doesn't necessarily imply good design or installation of whole systems or any work that isn't related to gas.
- **Heating and Ventilating Contractors' Association**

157

COMPARATIVE COSTS FOR SPACE HEATING
PER KILOWATT HOUR

Electricity
Fan heater/bar fire (general tariff) 6.21p
Storage heater (Economy 7 tariff) 2.75p [1]
Gas
Wall heater 1.89p
Radiant/convector heater 2.09p
Central heating radiator (conventional boiler) 1.94p
Central heating radiator (condensing boiler) 1.60p
LPG
Butane room heater 5.54p
Central heating radiator (propane conventional boiler) 3.36p
Central heating radiator (propane condensing boiler) 2.76p
Oil
Central heating radiator 1.61p
Paraffin
Room heater 3.62p
Solid fuel
Central heating radiator (anthracite grains) 2.31p
Open fire (coal group b) 4.02p
Room heater (Sunbrite) 2.30p

[1] Includes 10 per cent peak rate allowance
Running costs are based on average efficiencies for modern equipment or systems. Prices were correct as at September 1989

(HVCA) HVCA operate a home heating linkline which will provide the names and addresses of two HVCA-registered installers in your area. Tel: 0345 581158
- **Institute of Plumbing**
- **National Association of Plumbing, Heating and Mechanical Service Contractors**
- **Scottish and Northern Ireland Plumbing Employers Federation**
- **Solid Fuel Advisory Service**

What to look for in a design quote

Insist on a detailed design specification and ask questions if you're unsure whether or not the system is properly tailored to your needs. The installer should advise you on different levels of sophistication of controls; if he's got no ideas beyond a timer and simple room thermostat, don't use him! A good design quote should cover the following points:

- **Insulation** What assumptions have been made about the level of insulation or whether you need insulation work to be carried out?
- **Design temperatures** These are the internal temperatures your central heating will be designed to give when it's cold outside (for example, 21°C (70°F) in the living-room when it's −1°C (30°F) outside)
- **Air changes** The number of air changes allowed for in each room should be quoted
- **Equipment** The type and location of the boiler, pump, radiators, controls and hot water storage cylinder (if required). Whether the pipework will be concealed or surface-mounted
- **Standards** What regulations and standards the installer will work to. These should include Gas Safety, Wiring and Building Regulations, water by-laws, British Standards and trade association codes of good practice
- **Responsibility** Make sure you know whose responsibility it is to paint radiators, lift and replace floor coverings, make good where pipes are run through walls, and dispose of scrap materials or old equipment if you don't want it.
- **Before the installer comes to visit**, make sure you've considered: what style you would prefer radiators to be, where you want them, and whether it matters how high or long each one is; where you're prepared to have the boiler; whether you propose to make any other changes to the house, such as fitting new windows or adding an extension.

Choosing a new boiler

A central heating boiler should last at least 15 years, so if your boiler needs replacing it's likely to be a model that was designed and installed in the 1960s or early 1970s. You can expect a new boiler to be more compact and convenient and, if it's correctly sized, to use significantly less fuel to heat your house and water. You'll be able to either cut your fuel bills or heat your home to a higher standard for the same money.

However, replacing a boiler that's still working well is unlikely to pay immediate financial dividends (taking into account fuel bill savings compared to the cost of a new boiler); you should certainly do cheaper improvements like increasing insulation or adding controls first.

If you're choosing a new boiler you'll need to decide:

Do I want to change fuel?
A new boiler presents an opportunity to change fuels. At the moment oil is one of the cheapest fuels, but predicting prices is tricky. If you're in an area that can't get mains gas, it may be worth finding out if there are any plans to extend the service nearby.

For more details on choosing a fuel and fuel prices, see page 156.

Do I want to move the boiler?
Modern boilers are generally much smaller than they used to be, and you should be able to replace a floor-standing unit with a wall-hung version that takes up no more room than a kitchen cabinet. If you have an open fire and you need the space in the kitchen, you could have a back boiler behind the fire. If you're not happy with the current site or size of your boiler, ask your installer what options are open – he'll need to look at the possibilities for siting the flue.

Do I need the same size of boiler?
It's common for old boilers to be oversized – simply too big

for your heating needs. It may have been put in to cope with a larger heat requirement when the house was less well insulated or the heating less well controlled, or a careful calculation of the heat loss from your home may not have been done. A good central heating installer will choose a boiler that is closely sized to your needs – making it more efficient.

Types of boiler

For most people choosing a boiler will mean finding a suitable model from the conventional floor-standing or wall-hung gas-fired boilers available. But over the last few years several new types of boiler, designed to save energy or be more convenient, have also become available.

Condensing boilers

Condensing boilers cost more than conventional boilers, but they are much more efficient. A condensing boiler professionally installed will cost from around £1200 compared to from around £700 for a conventional boiler, though prices will vary according to circumstances. A condensing boiler should achieve average efficiencies over a year of 85 per cent or more (that is, 85 per cent of the possible heat in the gas burnt by the boiler is made available to heat the house and water) – the highest for any fossil-fuel boiler. Unlike a conventional gas boiler, efficiencies remain high even when working at low-level output – in the summer, for example.

The Chart on page 162 shows the savings that can be made by installing a condensing boiler for three types of houses in the Midlands. Heating costs from a system with a conventional floor-standing 15- to 20-year-old gas-fired boiler, typically oversized and in poor condition, are compared to a modern wall-hung conventional gas boiler that is correctly sized and to a correctly sized condensing boiler. Figures are based on recent research carried out by Ofgas – the Office of Gas Supply. They are consistent with independent research carried out elsewhere – for example, by the Building Research Energy Conservation Support Unit. Efficiencies for the three types of boiler over a year are around 55 per cent, 70 per cent and 85 per cent respectively.

COMPARATIVE HEATING COSTS

House	Old	Modern	Condensing
		Boiler type	
Detached	£887.63	£705.27	£587.22
Semi-detached	£563.86	£451.18	£378.11
Traditional terraced	£499.46	£400.52	£336.44

Heating costs for conventional and condensing gas boilers, based on houses in the Midlands heated to 21°C (70°F) for a total of seven hours a day.

How they work

Condensing boilers work in a fairly similar way to a conventional boiler but have either a second or a very large heat exchanger which recovers much of the heat that is usually lost in the flue gases. This heat is transferred to the water in the boiler that's returning from the radiators. Giving up this extra heat cools the flue gases below the 'dew point', and moisture held in them condenses out and is collected and drained away.

Condensing boilers currently available have fan-assisted balanced flues – because the flue gases are no longer warm enough to float away and need a fan to aid dispersal. As with many modern boilers, spark ignition replaces the wasteful pilot light.

Are they suitable for any system?

Most condensing boilers are designed for fully pumped gas central heating systems, though a few are made to run on LPG – propane or oil. They are not suited to systems that are controlled solely by TRVs without an overriding room thermostat. This is because for the boiler to work in condensing mode the return temperature of the radiator water must be kept down – ideally lower than the dew point of 54°C (129°F). It helps to keep the return temperature down if your existing radiators are oversized, but it wouldn't be worth changing to new oversized ones. However, a condensing boiler will always be more efficient than a conventional boiler, even if the return temperature of the water can't be kept as low as 54°C (129°F).

What about installation?
Condensing boilers are slightly more complicated to install because a drain needs to be fitted. Careful consideration of the site of the flue terminal is important because when it's operating in condensing mode a slight mist – called 'pluming' – is expelled from the terminal. If an inhibitor is used to prevent corrosion it must be one suitable for use with the aluminium heat exchangers in condensing boilers. Installation will be more expensive if you need to convert an existing gravity system to a fully pumped one – but this in itself may improve efficiency.

Verdict
If you're replacing a gas boiler, it's definitely worth considering a condensing boiler – the extra cost will usually be recouped in two or three years and perhaps more quickly in a large house. But a condensing boiler won't suit every system.

Combination boilers
Usually called a combi, this type of gas boiler has become very popular.

How they work
Combis work by combining the function of an 'instantaneous' water heater and central heating boiler in one unit.

The hot water for taps is heated and supplied 'instantaneously' – saving energy because you're not keeping a whole cylinder of water hot. The water for the radiators is heated in a similar way to a conventional boiler.

Are they suitable for any system?

Because a combi heats and delivers hot water direct and has no storage capacity, the flow of hot water will be slower – though you'll never run out. So if your demand for hot water is high – you tend to have baths rather than showers or there are four or more in the family – and it would cause inconvenience to have to wait, a combi won't be suitable.

Generally, a combi will provide a hot water supply to only one tap at a time, and it will not provide heat to the radiators while it is providing hot water. A bath will take about 15 minutes to run.

What about installation?

Because it's fed direct from the mains, a combi eliminates the need for a hot water storage cylinder and (because the boiler is 'sealed' with its own expansion vessel) you also don't need a feed and expansion cistern and all the pipework that goes with it. Combis tend to be popular with installers because they are so easy to fit, with less pipework and disruption than a conventional boiler.

But if you're having a combi installed, ask the installer what provision he has made for water treatment and whether the mains water pressure at peak use times is sufficient. Water treatment is important because the fresh water flowing through the combi makes it vulnerable to scale formation in hard water areas. Combi manufacturers will be able to advise you on how hard the water is in your area and what you can do about it. Maintenance costs for combis may be higher than for conventional boilers.

Verdict

Worth considering, especially for smaller homes with little storage space.

Coalflow boilers

These are designed to improve the convenience of using a solid-fuel boiler by reducing the need for refuelling and removing ashes. Coal is fed mechanically into the boiler from a hopper at the side. It burns ordinary coal smokelessly and can run for up to a week without refuelling.

Economy 7 boilers

These are sold primarily to replace oil boilers. They use off-peak electricity either to heat up water in a large storage tank to near boiling point, which is then pumped around the system, after mixing with cooler water, during the day; or, in another design, the heat is stored in a solid core similar to a storage heater, the water being heated by a heat exchanger. This overcomes to some extent the problem of space required by the water-only type.

=12=

INSULATING YOUR HOME

Efficient insulation can reduce the amount of energy consumed in a normal house by over a third. And a house designed from the start to minimise its demand for energy can use less than half the energy consumed in a conventional home of the same size. So there's great potential for reducing your demand for energy by improved insulation. And because all energy has to be paid for, there's scope for saving money, too.

In a poorly insulated house with a total annual fuel bill of £900, about £600 goes in space heating and water heating. With efficient insulation, space and water heating for the same house could cost as little as £235 – less than half.

The energy balance

We use energy to keep us warm and comfortable and to provide hot water. A large part of that energy is provided by the heating system, but there are other sources of energy which help to warm the house:

- almost all the energy used by lighting and by household appliances – TVs, hair driers, kettles, irons and so on – ends up as heat
- the energy used for cooking
- the heat from our bodies
- sunlight coming through the windows.

Insulating against high fuel bills

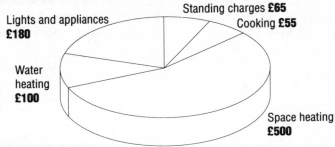

Standing charges **£65**
Cooking **£55**
Lights and appliances **£180**
Water heating **£100**
Space heating **£500**

Poorly insulated house

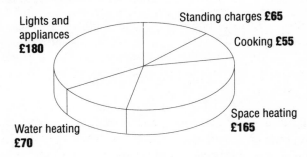

Standing charges **£65**
Cooking **£55**
Lights and appliances **£180**
Water heating **£70**
Space heating **£165**

Well insulated house

Heat goes through the walls, windows, roof and floor, up the flue or chimney, as draughts or ventilation, and as hot water down the drain. The inside temperature depends on the balance between the energy coming in and the energy being lost. By reducing the rate at which heat escapes, you can maintain the same inside temperature while consuming less energy.

The rate at which heat can pass through the materials of a house depends on:

- the types of materials
- the thickness and area of each material
- the temperature difference between the outside and inside.

167

Different types of materials lose heat at different rates. Architects and builders express these differences as U-values. The U-value of a material is the amount of heat, in watts, which will pass through one square metre of the

U-VALUES FOR BUILDING MATERIALS

			U-val W/m²/°C
solid walls	brick	105mm thick	3.1
		220mm thick	2.1
cavity walls	brick outside and inside	uninsulated	1.6
		with cavity insulation	0.5
	brick outside, block inside	uninsulated	1.0
		with cavity insulation	0.4
doors	in external walls	wood	2.5
		glazed	3.2
windows	single glazed	metal	5.6
		wood	3.5–4.5
	double glazed	wood or plastic	2.5
		metal frames, without thermal break	4.3
		metal frames, with thermal break	3.7
roofs	pitched	uninsulated	2.7
		50mm insulation	0.6
		100mm insulation	0.4
		200mm insulation	0.2
floors	solid	uninsulated	0.7–1.5
		insulated	0.3–0.8
	suspended	uninsulated	1.4
		insulated	0.7

material for each degree of temperature difference between the inside and outside. The Table opposite shows typical U-values for a range of building materials. The smaller the U-value, the lower the rate of heat loss. You can see from the Table where insulation can have a substantial effect. For example, insulation reduces the U-value of a typical cavity wall by more than half, and loft insulation can reduce the heat loss through a roof by over 90 per cent. But also note that double-glazed windows with metal frames have U-values similar to those for single-glazed wooden windows.

A strategy for effective insulation

There's a wide range of measures which you can take to improve the insulation of your home, and a wide range of costs associated with them. The benefits also vary, and the cheapest measures are not necessarily the most cost-effective, nor are the most expensive necessarily the least cost-effective. The best way to compare them is to look at the time it takes for each to pay for itself in terms of the energy saved. The Table on pages 170–2 lists a range of energy-saving ideas for home insulation. The approximate cost for a typical house is given, along with a rating reflecting the potential energy pay-off.

Insulation: how to go about it

Insulation costing less than £50

Fitting a jacket to a hot water cylinder
D-i-y rating: easy
Typical cost: £8 (d-i-y)
Typical annual saving: £16
Energy rating: ***
Insulating your hot water cylinder is one of the easiest and most cost-effective ways to reduce your energy consumption. Cylinder jackets are made in a range of standard sizes,

INSULATION IDEAS

			Typical cost /house	Typical annual saving	Rat
Insulation costing less than £50					
Hot water cylinder	Fit 80mm insulating jacket to uninsulated cylinder	d-i-y builder	£8 £20	£16	***
	or fit 80mm insulating jacket on top of factory-applied foam insulation	d-i-y builder	£8 £20	£3	**
Radiators	Put reflective foil behind radiators on outside walls	d-i-y builder	£10 £25	£4–£12	***
Windows	Fit draught-proofing	d-i-y builder	£30 £135	£25	***
Doors	Fit draught-proofing to exterior doors	d-i-y builder	£15 £50	£10	***
Suspended timber ground floor	Seal gap between skirting and floorboards with sealant	d-i-y builder	£14 £60	£8	***
	or seal gap between skirting and floor-boards with timber moulding	d-i-y builder	£35 £80	£8	**

based on the height of the tank from the base to the top of the dome, and on its diameter. Common sizes are 900mm x 450mm and 1050mm x 450mm. They should conform to British Standard 5615:1978 – don't buy one without the BS Kitemark.

			Typical cost /house	Typical annual saving	Rating
Insulation costing £50 to £200					
Roof	Lay 150mm loft insulation in uninsulated loft	d-i-y builder	£100 £230	£110	***
	or lay 150mm loft insulation instead of 100mm	d-i-y builder	£100 £230	£10	*
	or add 100mm extra loft insulation on top of existing 100mm insulation	d-i-y builder	£75 £205	£25	**
Windows	Add seasonal secondary double glazing	d-i-y	up to £50	£20	**
	or add fixed secondary double glazing (For professionally installed secondary double glazing, see next page)	d-i-y	£190	£20	*
Suspended timber ground floor	Take up floorboards, add 60mm of insulation supported by plastic netting between joists	d-i-y builder	£140 £375	£35	**

Most jackets consist of about eight segments drawn together by a string at the top and held in place by two or three belts. Smooth the jacket down over the cylinder, but do not press on it – if you crush it you'll reduce its insulating properties, so don't pull the belts too tight. Avoid gaps

171

		Typical cost /house	Typical annual saving	Ratir
Insulation costing more than £200				
Windows	Add professionally fitted secondary double glazing	£1250	£20	-
	or fit replacement double-glazed windows	£3500	£0–£40	-
Walls	Have cavity wall insulation installed	£350–£600	£100	**
Solid ground floor	Replace screed with 50mm polystyrene and moisture-resistant chipboard	£350	£40	*

Key
*** Good value: will pay for itself within two years
** A good investment: will pay for itself within five years
* A long-term investment: will pay for itself if you stay in the house for at least ten years
- Will save energy but is not cost-effective

between segments where heat could escape, but don't cover the cap or cables of an electric immersion heater.

If you need a new hot water cylinder, choose one with factory-applied foam insulation. Even these are worthwhile fitting with an extra insulating jacket, though.

As well as the cylinder, hot water pipes can be responsible for considerable heat loss. Foam plastic pipe insulation is cheap and easy to apply. It's split down one side so you can get it around the pipe, and the edges are treated with an adhesive. For extra security, use electrical insulation tape at intervals and especially near bends.

If you've already got an old cylinder jacket, it's probably

worth fitting a new one over the top. A double-insulated cylinder and lagged hot water pipes should still provide enough warmth in an airing cupboard.

Fitting reflective foil behind radiators
D-i-y rating: fairly easy
Typical cost: up to £3 per radiator (d-i-y)
Typical annual saving: £1 to £3 per radiator
Energy rating: ***

For radiators which are fitted to outside walls, a significant part of the heat goes into the wall rather than directly into the room. To keep the heat in, you can buy specially made reflective panels fitted with adhesive pads; a 10m roll costs about £10. But ordinary kitchen foil can be equally effective, and at a considerably lower cost. Simply stick the reflective panel, or the foil, to the wall behind the radiator. It's not unsightly, since the radiator makes it almost invisible.

If you can lift the radiator off the wall, it's a straightforward job. If you can't, try smoothing down the reflective sheet using a small, long-handled roller designed for painting behind radiators.

Draught-proofing windows and doors
D-i-y rating: fairly easy, but can be time-consuming
Typical cost: £45 for an average house (d-i-y, £185 installed professionally)
Typical annual saving: around £35
Energy rating: *** d-i-y, * professional

Houses must have a certain amount of ventilation to prevent condensation and to provide oxygen for fuel-burning heaters, boilers and cookers. But too much ventilation means heat going to waste.

Fit draught-excluders to windows, doors and over the letter-box. There's a wide range of types in the shops. For windows there are excluders which fit between the window and the frame, so they're out of sight when the window is closed. But they won't work if the gaps are too big. Alternatively, there are types which fit on the inside of the frame, so

173

that the window closes against them. Many can be fitted with nothing more than scissors or a trimming knife to cut them to length. Some may need a fine-toothed saw.

For doors, around the sides and tops you can use the same draught-excluders as for windows. For the bottoms of doors there are more substantial under-door strips, or seals which attach to the face of the door. Some of these hinge up when the door is opened. For the more complicated types, you may need a plane, hacksaw, bradawl and screwdriver.

Sealing up a draughty floor

D-i-y rating: fairly easy
Typical cost: from around £14 (d-i-y, £60 done professionally)
Typical annual saving: around £8
Energy rating: *** d-i-y, * professional

Doors and windows aren't the only places where draughts can occur. Suspended timber ground floors can give rise to troublesome and wasteful draughts. Don't be tempted to block off the airbricks in the wall beneath the floor – these are essential for under-floor ventilation, to keep the floor dry and free from rot.

Gaps between floorboards can be filled with strips of wood or papier mâché. Gaps between the floor and the skirting-boards can be filled using a sealant, or eliminated by fitting scotia or quadrant moulding.

Don't neglect sealing up any gaps where pipes or cables pass through the floor. Hardboard fixed over the boards eliminates draughts and is in any case necessary before laying vinyl tiles or sheet flooring.

Don't overdo draught-proofing in rooms where there's a lot of water in the air – principally kitchens and bathrooms – or where there are fuel-burning heaters, boilers or cookers. If condensation is more of a problem after draught-proofing, fit ventilation you can control. Extractor fans in kitchens and bathrooms can make a big difference, and you can get 'trickle ventilators' which fit in the tops of the window-frames. But it may be enough to simply open a window for a few minutes each day; this will allow the moist air to escape.

Insulation costing £50 to £200

Insulating the loft

D-i-y rating: fairly easy, but unpleasant work
Typical cost: around £100 (d-i-y, £230 installed by builder)
Typical annual saving: around £110
Energy rating: ***

Without loft insulation, about a quarter of the heat in your house will escape through the roof. Efficient insulation can reduce the loss to less than five per cent.

A number of materials can be used for loft insulation, some in rolls, others as loose-fill granules. The most effective materials are mineral fibre blanket and blown mineral wool or cellulose fibre. For d-i-y, the best bet is mineral fibre roll, which is readily available; if you're starting from scratch, go for a thickness of 150mm. The fibres can cause irritation, so wear a mask and gloves, and keep your arms covered.

Start by sealing any gaps in the ceilings where pipes pass through – inside the airing cupboard, for instance. Put a board across the joists in the loft, to stand or kneel on. Unroll blanket insulation between the joists, cutting to length as you go. With loose-fill insulation, pour the contents of the bag between the joists and, provided that they're at least 150mm deep, level off with a piece of board. Take care not to damage wiring, and cover only as much of it as you have to – it mustn't get too hot. Don't insulate under water cisterns: they need a little heat from below to stop them freezing.

Take care not to block ventilation at the edges of the roof; ventilation in the loft is all the more important when it's efficiently insulated, to prevent condensation and consequent rot in the timbers. If you can see the underside of the tiles or slates of the roof, or daylight at the eaves, then you probably don't need additional ventilation. But if your roof is felted or boarded under the tiles, you must provide ventilation at the eaves. At its simplest, this can be 50mm holes at about 200mm intervals bored in the soffits – the

175

horizontal boards under the edges of the roof – with patches of chicken wire to keep birds out. Alternatively, you could fit purpose-made soffit ventilators or ventilated ridge-tiles.

If you're receiving income support, family credit or housing benefit you may qualify for a grant of 90 per cent of the cost of loft insulation from your local authority.

Fitting seasonal double glazing

D-i-y rating: easy
Typical cost: up to £50 for whole house
Typical annual saving: £20
Energy rating: **

Seasonal double glazing is temporary secondary glazing applied inside the windows for the winter. At its simplest it can be plastic film – similar to the film used to wrap food – which costs as little as £3 per square metre. For a more permanent solution, there's rigid plastic sheeting – from about £10 per square metre – which can be held in place with magnetised adhesive tape. This allows you to remove the plastic sheeting if you need to escape from a fire, for instance; the sheeting is also easy to remove altogether in the warmer months.

Fitting d-i-y secondary double glazing

D-i-y rating: can be complicated
Typical cost: £190 for whole house
Typical annual saving: £20
Energy rating: *

You'll find a range of d-i-y secondary double-glazing systems in the shops, using either glass or plastic sheet. Plastic glazing has the advantage of being light and relatively easy to cut, whereas glass is more scratch-resistant.

Hinged secondary double glazing usually consists of a rigid frame in which the glass or plastic is mounted, incorporating a seal where the secondary frame meets the original window-frame. On fixed windows the secondary glazing panels can be fixed all round with clips or slotted channel. On opening casement windows the secondary glazing is hinged on one side and held in place by clips on the other sides. If you want a greater separation between the original glass and the new secondary glazing – for better sound insulation – the secondary glazing can be mounted on a sub-frame of battens fixed inside the window reveal.

Sliding secondary glazing also has a frame in which the glass or plastic is mounted, but then a second frame with tracks in which the glazing panels can slide. Normally, they're designed for the panels to slide horizontally, though vertical sliding types are available for sash windows. Sliding types tend to be rather more expensive than hinged systems, and can be complicated to fit.

Insulating a suspended timber ground floor
D-i-y rating: difficult
Typical cost: £140 (d-i-y, £375 installed by builder)
Typical annual saving: £35
Energy rating: ** d-i-y, * builder
You can insulate a suspended timber ground floor using the same mineral fibre rolls that are used as loft insulation, or with expanded polystyrene 'bats'. The problem is that unless you have access from below, you'll have to take up the floorboards to do it. The normal technique is to support the insulating material on plastic garden netting stapled between the joists.

Insulation costing over £200

Professional double glazing
Typical cost: secondary double glazing £1250, replacement windows £3500
Typical annual saving: £0 to £40
Energy rating: -
Professionally installed double glazing can't be justified in terms of the potential reduction in your fuel bills, whatever the installers might say. But it can be a way of eliminating

draughts, of giving you more control over ventilation and can help with condensation problems.

If you do decide to have your windows replaced, make sure that the new frames are well insulated. Metal frames are not as energy-efficient as plastic or wooden ones, particularly if the metal frames don't incorporate a 'thermal break' between their inner and outer surfaces.

Cavity wall insulation

Typical cost: £350–£600
Typical annual saving: £100
Energy rating: **

As much as a third of the heat inside your house could be escaping through the walls. If the house has cavity walls – as most houses built since the 1930s do – then you may be able to reduce this waste of energy by having cavity wall insulation installed. Not all houses are suitable, though. Timber-framed homes should not have their cavities filled – their walls should already be well insulated – and there can be problems of damp penetration in walls exposed to heavy driving rain.

Cavity wall insulation is installed by drilling a series of holes in the brickwork and then pumping or blowing insulation into the cavity. Three types of material are used: polystyrene beads or granules, blown mineral fibre and urea-formaldehyde (UF) foam. Employ only a specialist installer. If you go for UF foam, the contractor should be registered with the British Standards Institution and be prepared to guarantee that the work will comply with BS5618. For other materials, ensure that the material used is approved by the British Board of Agrément (address on page 216). Check that the firm will give you a long-term guarantee, and that the guarantee can be transferred to future owners.

UF foam gives off formaldehyde fumes while it's curing. Although the amount of these fumes is small, they could cause an allergic reaction in some people. If anyone in the household reacts badly to everyday chemicals such as cosmetics, choose one of the other materials.

Insulating a solid floor
Typical cost: £350 (builder)
Typical annual saving: £40
Energy rating: *

Most newer houses are built on solid concrete floors. These generally consist of a layer of rubble, followed by about 100mm of concrete, a damp-proof membrane and a finishing 'screed' about 50mm thick. Few existing solid floors incorporate insulation.

Like many energy-saving ideas, floor insulation is a good deal easier to put in when building from scratch. Nevertheless, it is feasible to add insulation to an existing solid floor, though it's a messy and very disruptive business. The costs and savings are based on the idea of replacing an existing floor screed with a laminate of 50mm extruded polystyrene and moisture-resistant chipboard finished with a floor sealer.

Some other energy-saving ideas
- Fit roller blinds to windows.
- Fit window-shutters made from hardboard and wooden battens as a cheap alternative to double glazing.
- Fit thermostatic valves to the radiators (see page 154).
- Turn the room thermostat down a little – just a degree or two will produce a noticeable saving on your bills.
- Replace your gas boiler with a more efficient condensing boiler (see page 161).
- If you have off-peak electric storage radiators, add an outside temperature sensor to control the overnight charge.
- When rebuilding or extending your home, design from the outset with energy-efficiency in mind.
- When you're thinking of buying a house, make a point of asking about its energy-saving features.

=13=

CONSUMING POWER

There is a lot of variation in the amount of energy different appliances use and in how people use them. How much energy an appliance will consume depends on its wattage (that is, whether it has a powerful motor or heater) and how much it is used. Some appliances – refrigerators, fridge-freezers, freezers – are on all the time; others – such as cookers, washing machines, tumble driers and televisions – are used frequently, perhaps every day; and some are used only every so often – hair driers and food mixers, for instance.

If you want to cut down the energy consumption of your household, you should buy the most energy-efficient appliance that suits your needs. For some of the big energy users like refrigeration equipment, *Which?* tests have shown that there can be significant differences in the energy consumption between different brands and models of the same appliance. Say an energy-efficient fridge, washing machine, tumble drier or dishwasher costs around £30 more than for the least efficient. In the first year you would get this back because of lower running costs, and save nearly 600 kilowatt hours (kwh) of energy. Assuming the appliance lasts for around 10 years, you could save around 6000kwh of energy and about £370 in running costs (based on current electricity costs).

In the following sections we look at the kind of differences you can expect between different types and models of appliances. For refrigeration equipment, washing machines, tumble driers and dishwashers, we've included a shortlist of models from recent *Which?* reports which have performed well for energy-efficiency (and at least reasonably well for performance). The shortlists are in alphabetical order. Of course, when choosing a domestic appliance, many factors are involved.

You should also use your appliances in the most energy-efficient way. With the most power-hungry appliances like cookers, washing machines, tumble driers and dish-washers, their energy consumption is affected by how they are used and how often. Even with refrigeration equipment, things like where you put it have an effect on consumption.

How much power?

Electricity

Electrical power is measured in watts (w). The rating plate of an appliance will tell you in watts the power consumption when the machine is on. The amount of electricity used for doing a job is measured in units, based on power consumption in an hour. You could give this figure in watts per hour, but it is generally given in kilowatts per hour (kwh), with one kwh being 1000w. So a 1000-watt appliance working for one hour will use one unit, whereas a 100-watt appliance will work for ten hours for one unit. If you divide the wattage figure on the rating plate by 1000, you get the number of kwh the machine uses for each hour it is on.

The Table on pages 182–3 gives a guide to the energy consumption for some common domestic appliances. Use this as a general guide; the power rating of appliances will vary from manufacturer to manufacturer, and how people use them will also have an effect.

Also, some appliances, such as irons and ovens, do not work continuously – they're fitted with a thermostat which switches off the appliance once the required temperature has been reached, switching on again when it drops below the setting.

Appliance usage	Guide to energy
Blender	500 pints of soup for 1kwh
Carving knife (electric)	Carve a joint every weekend for 4 years for 1kwh
Coffee percolator	Make 75 cups of coffee for 1kwh
Cooker: radiant ring (hob)	Cook chicken stew for 4 for 0.5kwh
Grill	Grill 0.5kg (1lb) of sausages for less than 0.5kwh
Contact infra-red grill	Cook 7 pork chops for 1kwh
Deep-fat fryer	Fry 2.25kg (5lb) of chips for 1kwh
Dishwasher	Wash 12 place settings (normal programme) for 2kwh
Food mixer	Mix 1 cake a week for a year for 1kwh
Freezer (upright or chest)	Run for 24 hours for 1 to 2kwh
Fridge-freezer	Run for 24 hours for 1 to 2.5kwh
Hair drier	Dry hair twice for 1 unit
Iron	More than 2 hours ironing for 1kwh
Kettle	Boil 12 pints water for 1kwh
Microwave oven	Cook 8 chicken pieces or 2 joints of beef for 1kwh

Radio	24 hours a day for 1 week for 1kwh
Refrigerator	24 hours for 0.5 to 1kwh
Slow cooker	Cook 1 pint of rice pudding for less than 1kwh. Cook a chicken casserole for 1.33kwh
Stereo system	Listen for 8 to 10 hours for 1kwh
TV (22-inch colour)	6 to 9 hours' viewing for 1kwh
Tumble drier	Dry a 4.5kg (10lb) load for 2 to 3kwh
Vacuum cleaner: cylinder	1 hour vacuuming for 1kwh
upright	2 hours' vacuuming for 1kwh
Washing machine: front-loader/ top-loader (drum type)	Wash 4.5kg (10lb) load at 95°C (203°F) for 2 to 3kwh. Wash 2kg (4½lb) load at 50°C (122°F) for 0.5 to 1kwh
Waste disposal unit	Grind 22.5kg (50lb) of rubbish for 1kwh

Gas

The amount of energy used by gas appliances is measured in therms. One therm is the equivalent of about 29kwh of electricity. Working out how much gas an appliance uses is not as simple as with electricity as you have to know not only its 'rating' but also how efficiently it burns the gas. The

Table below gives a general guide to how much use you get for one therm.

Appliance		Amount of use for 1 therm of gas
Cooker: hot-plate	high setting	8 hours
	low setting	100 hours
grill	high setting	7–8 hours
oven	Gas mark 2	36 hours
	Gas mark 5	26 hours
	Gas mark 7	20 hours
Fridge		7 days
Fridge-freezer		5 days

Fridges, fridge-freezers and freezers

Fridges and freezers consume energy at a slow rate compared with other large appliances, but because they are plugged in 24 hours a day, 365 days a year, the energy used mounts up. Depending on type, efficiency and room conditions, fridges use around 0.5 to 1kwh of electricity per day; fridge-freezers and freezers use around 1 to 2.5kwh per day.

Which? tests have shown that the energy consumption of refrigeration equipment can vary greatly, and some models can use over twice as much energy as others of similar storage capacity.

Which? shortlist

Fridges

Worktop-height larder
Bosch KTR1541 £220
Eurotech LR1001 [1] £130

Frigidaire R1590	£140
Philips Whirlpool ARG716 [2]	£180
Zanussi ZR60/L	£240

[1] Exclusive to Iceland/Bejam stores
[2] Tested as Philips ARG193/PH

Worktop-height plus frozen food compartment

| Bauknecht KDC1533 | £230 |
| Frigidaire R1516 | £100 |

Tall larder

Asko Polar KS3501	£370
Bosch KSR2512	£325
Proline R111 [3]	£220

[3] Exclusive to Comet. Scandinova KS315-47 (£230) is similar

Fridge-freezers

One-control

Bosch KSV2621	£335
Bosch KSV3121	£370
Bosch KSV4312	£475

Two-control

| Bosch KGE3434 [4] | £480 |
| Scandinova RF7054 [5] | £350 |

[4] Tested as Bosch KGE3433
[5] Exclusive to Comet. Scandinova KF350-47 is similar

Freezers

Large chest

Electrolux TC1160	£230
Liebherr Comfort GT2602	£310
Philips Whirlpool AFB570	£250
Zanussi CF132	£240

[6] Tested as Philips 570/PH

Small chest
Zanussi CF64 £199

Upright
AEG Arctis 1415GS £270
Bosch GSD1440 [7] £345

[7] Tested as Bosch GSD143

Energy-saving tips

- When buying a new fridge or freezer, work out what size you need. Buying one that is too big is wasteful of energy; both fridges and freezers operate best when at least three-quarters full.
- Think carefully where you put the fridge or freezer. The kitchen is the obvious place, but remember the warmer the surrounding temperature the more energy it will use.
- Avoid placing it next to a source of heat, such as a cooker or central heating boiler. If you can't avoid it, leave as much space as possible between the two (a gap of at least 2.5 centimetres/1 inch).
- Avoid putting it near a window in direct sunlight.
- Open the door as seldom as possible, and don't leave it ajar – otherwise cold air escapes and the fridge has to work harder to get back to its correct temperature.
- Let hot cooked food cool down before putting it into the fridge or freezer.
- Defrost regularly.
- Check doorseals by closing the door on a piece of paper. If it isn't gripped tightly and slides easily, buy new seals.

Washing machines

With laundry appliances (and dishwashers) how much energy you use largely depends on how often you use the machine and what programmes you use.

A few washing machines are cold fill, which means they

186

heat up inside the machine all the water needed for a wash. Others use some hot water from the house supply. If you include the energy to heat the water outside the machine by electricity, most machines use around 2 to 3kwh to wash a 4.5-kilogramme (10-pound) cotton load at 95°C (203°F). For a 2-kilogramme (4½-pound) synthetics load at 50°C (122°F), they use around 0.5 to 1kwh.

In recent *Which?* tests there have been a couple of machines which use over twice as much energy on some programmes as others. Tub-type top-loaders use a lot more water per wash, and consequently they consume much more energy than front-loaders or drum-type top-loaders.

Water and detergent

All washing machines use a lot of water, but there are big differences between models. Also, some machines use detergent more efficiently than others, either by the way they circulate the wash water or by special valves on the drainage system. Others pre-fill the machine with some water before the detergent is washed from the dispenser so not as much gets away. You can virtually eliminate wasteful detergent loss by using a 'wash-ball' with liquid detergent (there is one specially concentrated powder) and put it directly into the drum. If you use less detergent there will be much less of it for the waste system to deal with.

Washer-driers

These wash and dry in one drum. Because of the design compromise needed to make a washer-drier do the work of two machines, their tumble drying performance doesn't match that of a separate tumble drier. The spin speed of a washer-drier is particularly important – the faster the spin the more water is extracted from the clothes and the quicker they are to tumble dry (see Tumble Driers, page 188).

Also, most washer-driers are condenser-vented, which means they use much more water than conventional tumble driers.

Which? shortlist

Because a number of front-loaders in recent *Which?* tests were fairly similar in their energy consumption, the machines below are shortlisted because they were among the most efficient. For water consumption, all the machines listed performed reasonably well (apart from the two washer-driers, which tended to use more). However, the Hoover condenser washer-drier was one of the machines which used least water on its drying cycle.

Front-loaders
AEG Lavamat 981 Sensortronic	£450
Zanussi FJ1023	£410

Top-loader
Philips Whirlpool 1100 AWG013	£440

Washer-driers
Hoover A8548/50 (air-vented)	£380
Hoover A8552/54 (condenser-vented)	£440

Energy-saving tips

- Wait until you have a full load of clothes before using the machine. If you can't do this, use a half-load option, if there is one, which may save some energy and will cut down on the water and detergent used.
- Experiment with a lower temperature programme. This is the best way to save energy, especially when your washing isn't all that grubby. A hot wash (95°C/203°F) is not necessary for most washing and uses a lot of energy.
- Try to pre-treat stains before washing. Soak clothes in detergent and water rather than using the pre-wash cycle on your machine.

Tumble driers

The most energy-efficient way to dry washing is to hang it

out on a clothes line. If you don't have access to an outside line and don't fancy using an indoor clothes horse, then you may decide a tumble drier is your only option. A fast-spinning washing machine will cut down the amount of energy needed to tumble dry washing. Tumble drying a four-kilogramme (9-pound) load of mixed cottons spun at 500 rev/min will use twice as much energy as the same load spun at 1000 rev/min. Generally, the faster a washing machine spins, the drier your washing is likely to be at the end. But *Which?* tests have shown that this is not the whole story: the difference between 850 rev/min and 1100 rev/min is not so critical, but some machines which claim 1000 rev/min and higher maximum spin speeds never actually spin at those speeds for very long, while some machines with lower maximum spin speeds spin for a longer time and thus achieve good spinning efficiency.

Typically, a tumble drier uses 2.25kwh to dry 4.5 kilogrammes (10 pounds) of cottons, 3.6kwh to dry the same weight of towelling and 1kwh to dry 2 kilogrammes (4½ pounds) of polycottons (assuming the cottons and towelling are spun at 1000 rev/min and polycottons at 500 rev/min).

Which? shortlist

Because many models are similar for energy consumption, we've picked out models which were both among the most efficient and recommended in *Which?* for performance, convenience and other advantages.

Creda Compact	
Reversair 37324 [1]	£110
Creda Sensair	
37449 [2]	£190
Zanussi TD101	£145

[1] Creda 37325 (£110) and Electra 37288 (£100) are similar but not reverse-tumble
[2] Electra 37469 (£175) is similar

189

Energy-saving tips

- Dry as much as possible on an outdoor line.
- Wait until you have a full load of clothes before drying.
- Make sure you spin as much water as possible out of the clothes before tumble drying.
- Don't over-dry the clothes — it can set creases in and make ironing more difficult.
- Regularly clean the filter.

Dishwashers

A full-size dishwasher (taking 12 place settings) uses around 2kwh on a 'normal' (hot) programme. If your dishes are lightly soiled, try using an economy wash – although *Which?* tests have shown that only a few machines have economy programmes which combine good performance with energy saving.

Slim-line or table-top dishwashers (taking between four and seven place settings) are also available and are worth considering for smaller households.

Dishwashers also vary in the amount of water they use – see Chapter 8, page 108.

Which? shortlist

Full-size

AEG Favorit 667	£400
Indesit D320BG	£220
Philips Whirlpool ADG664	£380

Slim

Bosch SPS5121	£330

Energy-saving tips

The most energy-efficient way to use a dishwasher is to run it only when it is fully loaded. Make sure you scrape dishes clean before putting them in the dishwasher. Use the rinse programme on items in the dishwasher until you have a full load.

Cookers

Cooking uses a lot of energy, but individual lifestyles, eating habits and the way a cooker is used affect energy consumption. The energy used varies depending on things like the type of food eaten; the number of meals people eat out or how often they buy ready-cooked food (shifting consumption from the domestic to the commercial sector); if the family eats together or individually at different times.

Gas or electricity?

Electric cookers are cheaper to buy than gas cookers with similar features. However, *Which?* tests have shown that electric cookers cost about three times as much to use as gas cookers. In particular, electric grills cost more to use than gas, because of their longer warm-up time.

Energy-saving tips

- Make maximum use of the oven by cooking several dishes at once whenever possible.
- For small portions, consider using the grill or hob instead of the oven. Some cookers have half-grills which let you heat only part of the grill. And some have dual-element rings on the hob which let you heat the whole ring or the central area only – useful and economical for small pans.
- If you have an electric grill, it is more energy-efficient to use an electric toaster to make toast.
- Some cookers have a second oven which is smaller than the main oven, but these are not necessarily quicker to heat up or cheaper to use than the main oven: second ovens on older cookers are usually less well insulated.

191

- Make sure the pan you use is the same size as the electric ring. With gas, make sure the flames are under the saucepan, not licking up the sides and wasting heat.
- Steam fast-cooking vegetables over a pan of slower-cooking ones. Cook green vegetables in very little water. Cut food into small pieces so it cooks more quickly.
- Use a pressure cooker for things that take a long time to cook, such as soups, stews or dried pulses.
- Make sure that saucepan lids fit tightly.

Microwave cooking

By speeding up cooking times, a microwave oven can cut energy consumption. If you have an electric cooker, a microwave will probably use about a third as much energy across a range of cooking; savings will be particularly great on things you would have cooked in an electric oven or under an electric grill. Comparisons with a gas cooker are not as clear-cut, particularly for things usually cooked on a gas hob, such as fish, beefburgers and carrots, where cooking times may be similar. Remember also that with a conventional cooker you can usually cook more items at the same time. Try not to use too much disposable packing, plastic wrap or paper towel when using your microwave.

=14=

APPLYING THE PRESSURE

Doing your bit for the environment by buying and using goods with care needn't be hard, but is it enough? Though you may decide to 'go green', your neighbours and friends may not. What about the enormous issues which you can't affect on your own, like industries and governments, even whole cultures of people who may be doing things to harm the environment? You may also find your attempts to be greener frustrated – suggesting you try to use the car less is all very well but not of much practical help if you live somewhere where there is no bus or train service.

There are some issues that really do need to be tackled at a national or even international level. The thought of influencing governments and large multinational companies can seem daunting, but there are environmental pressure groups whose aims are to do just that and which have achieved a great deal over the last few years.

Generally speaking, pressure groups use the individual's support to demonstrate that there is demand for the changes they want. Money raised from membership fees and other sources pays for offices and staff, enables pressure groups to build up expertise, carry out research and make people aware of what is going on. Groups provide members with information and advice, give them the opportunity to learn more about problems and do practical work.

If you do want to extend your influence further afield, get practically involved in some sort of conservation work or want to find out more about the issues raised by this book, your best bet is probably to join one of the many organisations that campaign on environmental issues. If you're interested in joining any of the organisations, write to the address given at the top of the entry.

The Ark Environmental Foundation
498–500 Harrow Road, London W9 3QA
081-968 6780

Ark is a relative newcomer to the scene, and its slogan – Fighting for Life on Earth – shows the very wide range of international issues it is concerned about – food safety, pollution of rivers and oceans, acid rain, and an end to nuclear power and weapons, to mention just a few.

Strictly speaking, Ark isn't a pressure group as it doesn't lobby industry or government at a national level. Its work is aimed at individuals – at increasing awareness of environmental issues and helping people to live life in a way that is less harmful to the environment. Money raised from membership fees, donations and other sources pays for publicity, educational materials and helps support Ark's local groups.

Ark believes that people who join want to help change the environment, and it suggests potential members try to follow six rules: to use unleaded petrol; to use household products that don't harm the environment; to save energy; to recycle household rubbish; to eat less meat and animal fats; to insist on organically grown fruit and vegetables.

Ark produces a range of cleaning products designed with the environment in mind. The products are available by mail order and in some supermarkets – profits are used to help fund Ark's work.

By joining one of Ark's local groups you can get involved in work in your own area, for example by setting up recycling schemes for glass and cans. The groups identify and encourage local shops to sell organic produce, organise campaigns, raise money and tell people about Ark and its work.

Membership facts
Number of members 10,000
Membership fees Individuals £12; family £17; unwaged, retired or under-18 £5
Magazine/newsletter Ark Times, black-and-white tabloid-style newsletter printed quarterly on recycled paper
Members also get Ark sticker and badge, Ark's manifesto
Mail order A small range of goods is available

The Association for the Protection of Rural Scotland (APRS)
14a Napier Road, Edinburgh EH10 5AY
031-229 1898
Involved in most aspects of Scotland's rural life, APRS keeps a watchful eye on planning applications, is consulted about fish-farming leases and produces reports on environmental issues that particularly affect Scotland – acid rain and the lochs, waste disposal, protection of Green Belts and the destruction of natural habitats for golf courses and housing.

Membership of APRS includes 1000 individuals and over 130 organisations and associations, including Community Councils, the National Trust for Scotland, the Ramblers' Association and a few major companies. Meetings of the APRS council are open to members, and guest speakers are often invited. If relevant, the meetings are open to the public.

Money to fund its work and pay for publicity and educational materials comes from subscriptions, sponsorship and donations. APRS also receives a small grant from the Scottish Office.

Membership facts
Number of members 1000 individuals
Membership fees Individual £8; retired and unwaged £5
Magazine/newsletter A newsletter is published two/three times a year, on recycled paper when possible
Members also get Invitation to APRS council meetings
Mail order None

The British Trust for Conservation Volunteers (BTCV)
36 St Mary's Street, Wallingford, Oxfordshire OX10 0EU
WALLINGFORD (0491) 39766

If you are looking for a practical, 'hands-on' way to help improve the environment, then BTCV's local groups are well worth considering. Around 60,000 volunteers get involved each year on projects such as footpath improvement, tree planting and repairing old stone walls. If you don't want to get involved on a regular basis, you could try a working weekend or holiday in the UK or overseas. Trained leaders are always on hand to give advice and to ensure that a professional job is done.

Money to fund BTCV's national work comes mainly from the Department of the Environment, the Nature Conservancy Council and the Countryside Commission. Fundraising and industrial sponsorship provides about a third of its income. The money raised funds national campaigns, pays for equipment and educational materials and supports a network of field officers to help local groups. The local groups are largely self-supporting. Tools, advice and some grants are available from BTCV nationally.

If you are already involved in practical conservation work in your own area, it could be worth getting your group involved with BTCV (affiliation costs £13). BTCV offers insurance, advice and practical help.

If you don't want to get involved in practical work, you can join BTCV to show your support and keep up to date with what is going on.

Scotland: If you live in Scotland it's the Scottish Conservation Projects Trust that carries out this type of work. You can contact them at Balallan House, 24 Allan Park, Stirling, Central FK8 2QG (telephone Stirling (0786) 79697).

Membership facts

Number of members   11,200 members of BTCV, plus around 60,000 volunteers working for local groups

Membership fees   Individual £10; family £14; unemployed, students and retired £5.50

Magazine/newsletter   *The Conserver*, a quarterly part-colour tabloid-style newsletter printed on recycled paper

Members also get Annual report, copies of the working holiday brochure *Natural Break*, discounts on BTCV books and products, plus information from the local groups

Mail order BTCV produces three catalogues: *Tools* (telephone DONCASTER (0302) 859522); *BTCV Goods*, including practical handbooks (telephone DONCASTER (0302) 859522); *Trees and Shrubs* (telephone BATH (0225) 874018)

Council for the Protection of Rural England (CPRE)

Warwick House, 25 Buckingham Palace Road,
London SW1W 0PP
071-976 6433

With its expertise in planning and countryside law, CPRE campaigns to protect Britain's towns, villages and countryside. It looks at how national and local government policies and legislation affect the landscape. Funded mainly by members' subscriptions and supporters' donations, CPRE carries out research, monitors legislation, lobbies government and helps raise public awareness.

CPRE has 43 county branches which, with help and advice from headquarters, tackle local problems. Members can help in many ways, by bringing specialist or organisational skills to assist their local group, by fund-raising or simply by becoming a member and supporting CPRE's work.

Membership facts

Number of members 45,000

Membership fees Individual £12; joint £16; under-25 £7

Magazine/newsletter *Countryside Campaigner*, a part-colour magazine, printed on recycled paper, issued three times a year

Members also get Annual report and information about local activities, meetings and campaigns

Mail order A catalogue is planned for Christmas 1990

Environmental Transport Association (ETA)

15a George Street, Croydon CR0 1LA
081-666 0445

197

Launched in 1990, the Environmental Transport Association helps people use transport in a more environmentally sound way. Members are encouraged to reduce their dependency on cars and to consider other options before getting into the driving seat. To help people use the car less, the ETA campaigns for better and cheaper public transport, safer footpaths and cycle-ways.

The ETA is self-financing. Money is raised through membership fees and the profits from a range of transport-related services offered to members. The car rescue service costs £38 if your car is under ten years old, £48 if it's between ten to twenty years old. Help is available if you break down at home or elsewhere, and the cost includes roadside assistance and towing. Other services offered by the ETA are insurance policies for environment-conscious motorists, holiday cover and a cycle insurance package with a no-claims bonus that includes cycle rescue cover.

Membership facts

Number of members   Membership of the ETA has just opened

Membership fees   Individual £16; family £28

Magazine/newsletter   *Going Green*, the ETA's bi-monthly magazine, is printed on recycled paper

Members also get   Cover for legal expenses and personal accidents, use of a 24-hour emergency helpline and access to the optional services described above

Mail order   None

The Food Commission

88 Old Street, London EC1V 9AR

071-253 9513

Campaigning for better health through safer, more nutritious food, the Food Commission (formerly the London Food Commission) researches into how additives, preservatives, pesticides, and ways of cooking and storing food affect its quality and our health. Through books, leaflets and *The Food Magazine*, it publicises food facts such as the nutritional content of fast food and the fruit content of 'fruit' juices. Using its research findings as evidence, it lobbies government, manufacturers and retailers to improve the quality of food and food labelling.

You can support the Food Commission's work and find out more about the issues by subscribing to *The Food Magazine*. An annual subscription costs £12.50, and the magazine is published quarterly.

Friends of the Earth
26–28 Underwood Street, London N1 7JQ
071-490 1555

With a rapidly expanding membership of over 180,000, Friends of the Earth campaigns to protect the environment by exposing those who are destroying it and putting pressure on governments, companies and others with the power to do something about it. Friends of the Earth's campaigns cover the whole spectrum of environmental problems – acid rain, the destruction of the ozone layer, traffic pollution, rainforests, pesticides in food and many others.

Money is raised through subscriptions, donations, appeals, by local groups and through trading. The funds are used to run campaigns, for publicity, educational and briefing materials and to pay for research. Friends of the Earth also organises boycotts and encourages members to write to their MPS, retailers, etc. on specific issues. The Arts for the Earth (TATE) organises major eye-catching events to raise money and promote Friends of the Earth's work.

Friends of the Earth welcomes all levels of support, from those who simply want to keep in touch with what's going on to those who want to get actively involved in the work of their local group. There are just under 300 local groups where you can take part in fund-raising initiatives, events to tell others about the problems Friends of the Earth is tackling and lobby local councils and MPS on national and local issues.

Membership facts

Number of members   180,000

Membership fees   Individual £12; Earth Action (14–23-year-olds) £5

Magazine/newsletter   Earth Matters, a quarterly part-colour magazine printed on recycled paper

Members also get   Information from local groups and campaign updates

Mail order   A catalogue offers a range of goods

Greenpeace

30–31 Islington Green, London N1 8XE

071-354 5100

Greenpeace is an international pressure group with over 340,000 paid-up supporters in the UK. In campaigning against the destruction of the natural world, Greenpeace lobbies governments and carries out and publishes scientific research. Greenpeace also organises direct action protests, which have included preventing whales from being harpooned and plugging pipes through which toxic waste flows into the sea. The protests are intended to be peaceful yet forceful ways to bring issues to the attention of the media, public and people empowered to bring change.

Greenpeace also organises boycotts and encourages members to write to companies and MPs on specific issues.

Greenpeace is currently campaigning against global warming, acid rain, the killing of dolphins and seals, the threat to wildlife through habitat destruction and hunting, nuclear power and weapons, the dumping of poisonous substances at sea and many other issues.

Money to run its campaigns comes from supporters' subscriptions, donations, fund-raising initiatives and from the sale of goods.

As a supporter, you can get involved with local groups whose main purpose is to raise funds. The groups also do some publicity work to drum up support for campaigns. You won't get involved in direct action – Greenpeace has its own activists who are specially trained for this work.

Membership facts
Number of members   342,000 (known as 'supporters')
Membership fees   Individual £12; family £17.50; unwaged £6
Magazine/newsletter   *Greenpeace News* is published quarterly
   on recycled paper
Members also get   Campaign updates
Mail order   A mail order catalogue offers a range of goods

The Henry Doubleday Research Association (HDRA)

Ryton Gardens, National Centre for Organic Gardening,
Ryton-on-Dunsmore, Coventry, West Midlands CV8 3LG
COVENTRY (0203) 303517

This is a society well worth finding out about if you're interested in organically grown food and gardening. HDRA has established a 20-acre garden (Ryton Gardens – entrance £2) where you can see organic gardening in action, taste the results at its café and buy everything from seeds to compost makers at the shop.

HDRA has a rapidly growing membership of 17,000. Money raised through membership, from donations, the gardening and organic wine catalogues and the charge made for entrance to the garden itself is used to fund campaigns, research and run the garden. HDRA believes that many more of the old varieties of fruit and vegetables (preserved in a seed bank) should be available for people to grow and is campaigning for a widening of the list of seeds that can be sold in the UK.

HDRA has also been campaigning to save the fruit research centre at Brogdale in Kent where old varieties of fruit trees and bushes are grown and compared.

HDRA takes an interest in wider issues and is helping the campaign to preserve Britain's diminishing peatlands by researching into alternatives to peat for the garden. HDRA's work extends into the Third World, with tree-planting schemes to help bring deserts back to life.

Members can get involved in a number of ways. Local groups are fairly independent, but activities can include practical help with organic gardening (including running allotments), talks and fund-raising. If you'd rather do something on your own, you can become a 'seed guardian' and nurture one of the rarer or older varieties by growing it in your own garden.

Members can take part in trials, growing a particular variety and perhaps weighing the crop to determine yield. Results are sent back to HDRA, together with information about such matters as your garden's aspect and soil type, and collated to give a national picture.

Membership facts

Number of members   17,000

Membership fees Individual £12; family £15; unwaged, retired, students and disabled £7

Magazine/newsletter The HDRA Newsletter, published quarterly

Members also get An introductory information pack and free entry to Ryton Gardens. HDRA gives free advice to members, who can telephone or write in with their queries

Mail order The gardening catalogue offers organically produced seeds, equipment and other goods, and there is a separate catalogue for organic wines

Marine Conservation Society

9b Gloucester Road, Ross-on-Wye,
Hereford & Worcester HR9 5BU
ROSS-ON-WYE (0989) 66017

The Marine Conservation Society is the only UK environmental group to campaign specifically for the protection of the seas and coast. The society works to prevent the pollution of the North Sea and Britain's beaches, to protect marine wildlife like seals and sharks and to stop the trade in fish from tropical seas and coral reefs. It campaigns by lobbying governments, by publishing scientific reports and by providing information and advice on marine conservation issues.

The organisation is funded through membership subscriptions, grants and donations and by money raised from the sale of books and other items. The society publishes *The Good Beach Guide* with information on pollution and cleanliness along Britain's coastline.

Through local groups the society runs courses, talks, walks and evening lectures to help people learn more about the sea and the plant and animal life it supports. Local groups can campaign to solve local problems of littering or pollution on beaches and can also take part in fund-raising initiatives. Members who prefer to be out in the air can help out with shore projects, and for keen divers there are opportunities to help with underwater research.

Membership facts

Number of members   5000 (individuals and organisations)

Membership fees   Individual £12; family £16; groups and institutions £30

Magazine/newsletter   Marine Conservation magazine, published quarterly, is printed on recycled paper

Members also get   A poster, sticker and membership card

Mail order   Members receive a mail order brochure offering publications, T-shirts, sweatshirts, posters and other gifts. Telephone ROSS-ON-WYE (0989) 62834 for a copy

National Society for Clean Air

136 North Street, Brighton, East Sussex BN1 1RG
BRIGHTON (0273) 26313

First set up in 1899 to campaign against pollution caused by burning coal, the National Society for Clean Air now campaigns on a wide range of issues – air pollution, waste disposal, noise, acid rain, the greenhouse effect and indoor air quality. The society publishes educational material, and lobbies and advises government on these issues. Membership is mainly confined to organisations, but individual members are also welcomed.

Membership facts

Number of members   1000 (includes companies/organisations)

Membership fees   £12.50

Magazine/newsletter   *Clean Air* is published quarterly on recycled paper

Members also get   Members' handbook, discounts on conferences and workshops

Mail order   A range of books, leaflets and educational materials is available

National Trust
36 Queen Anne's Gate, London SW1H 9AS

071-222 9251

Although better known for its work conserving England's stately homes, the National Trust also owns and protects 564,922 acres of coastline and countryside and the many species of animals, birds and plants living there. Money to buy and care for the Trust's property is raised through membership fees, appeals, donations, legacies, entrance charges, restaurants and shops at National Trust properties, mail order and many other initiatives.

Local groups organise visits to the Trust's properties, talks, films and fund-raising activities, and there are plenty of opportunities for volunteers.

Membership facts
Number of members   1,865,000

Membership fees   Individual £19 – additional members at the same address £10; family £34; under-23s £7.50

Magazine/newsletter   *The National Trust Magazine*, a glossy, full-colour magazine, is published three times a year on recycled paper

Members also get   A copy of the National Trust handbook with details of all its properties; free entry to the properties; a regional newsletter with each magazine giving details of local activities, and a copy of the Trust's annual report

Mail order   A catalogue offers books, preserves, sweaters, cards and many other goods

National Trust for Scotland
5 Charlotte Square, Edinburgh EH2 4DU
031-226 5922

Although similar to the National Trust, the National Trust for Scotland is a separate organisation. The Trust owns and manages over 100,000 acres, including mountainous areas, islands and large estates.

Voluntary help from members is welcome, and they can get involved with their local group – organising talks, special events and fund-raising activities.

The Trust runs Thistle Camps, where members can take part in practical conservation work such as repairing dykes and stone walls or building paths to keep visitors from damaging valuable land. Volunteers are also needed to act as guides at the Trust's properties.

Membership details
Number of members   200,000

Membership fees   Individual £15; family £24.50; under-23s £6; retired £7.50

Magazine/newsletter   *Heritage Scotland*, the Trust's quarterly magazine, is printed in colour on recycled paper

Members also get   A guide and free access to the Trust's properties and to those owned by the National Trust, plus details of local events

Mail order   Catalogue available

New Consumer
52 Elswick Road, Newcastle upon Tyne,
Tyne & Wear NE4 6JH
091-272 1148

New Consumer researches and publishes information about companies that enables people to see what they are up to – those involved in South Africa or which use animals to test their products, for instance – and to decide which companies and products they'd rather avoid. In a pocket-sized book it lists American companies' records for giving to charity, women's advancement, helping people from minorities, involvement in military contracts, and more. Work on a British version covering the top 130 companies in the consumer goods market is underway.

You can support New Consumer by subscribing to its magazine of the same name. The magazine reports New Consumer's findings and is published every three months. A year's subscription costs £12.

Parents for Safe Food

Britannia House, 1–11 Glenthorne Road, London W6 0LF
081-748 9898

Parents for Safe Food campaigns for safe, nutritious food. It believes people should not be exposed to unnecessary risk from food and that they should know about pesticides, additives and other chemicals so that they can choose to avoid them. Parents for Safe Food wants less risk from food adulteration and contamination and wants to encourage environmentally sound and healthier food production methods. It also wants a food and farming policy which takes consumers' views into account.

Set up in 1989 by people from the world of arts and entertainments to campaign for the withdrawal of Alar, a spray used on apples, Parents for Safe Food now has a general interest in all food safety and health matters. It commissions research into food-related issues, lobbies government and publicises its work to raise consumer awareness. It is not a membership organisation as such. Instead, people can become supporters and make contributions to its work. If you pay £5 or more you will receive information and briefings containing advice and information about safe food. A safe food kit and promotional T-shirts are available by mail order.

The Royal Society for Nature Conservation (RSNC)

The Green, Witham Park, Lincoln, Lincolnshire LN5 7JR
LINCOLN (0522) 544400

Britain's countryside is home to many wild flowers, birds and animals, but RSNC's survey of hedgerows, meadows, woods and heathlands showed that many unique habitats have been destroyed. RSNC works to protect those that remain, including countryside that is home to badgers, otters, barn owls, dormice, rare birds, orchids and other threatened species. Through its network of local Wildlife Trusts, the society owns or manages around 1800 nature reserves.

If you decide to join you will become a member of a local Wildlife Trust. Local Wildlife Trusts organise talks, walks and other events to help you find out more about the wildlife in your own area. A special club for children, Watch, is sponsored by RSNC, and for town and city dwellers there are 50 Urban Wildlife Groups linked to RSNC.

The society's main funding is from the Department of the Environment. Money is also obtained through fund-raising initiatives, corporate sponsorship and membership fees. It is used to buy and maintain nature reserves, fund national campaigns and provide publicity and educational materials.

Membership facts
Number of members   215,000
Membership fees   The cost of membership varies from group to group but ranges from £6–£12 per person; membership of Watch is £4
Magazine/newsletter   Members receive copies of *Natural World*, the society's full-colour magazine published three times a year
Members also get   Information from their local group, plus organised walks and talks on relevant topics
Mail order   A Christmas catalogue is available

The Royal Society for the Protection of Birds (RSPB)
The Lodge, Sandy, Bedfordshire SG19 2DL
SANDY (0767) 680551
The RSPB works to protect birds and the environment. Founded over 100 years ago to (successfully) campaign for a ban on importing plumes, the RSPB's current emphasis is on habitat conservation – particularly heathlands, estuaries, native pine woods, uplands, wetlands and peatlands.

The RSPB's money comes from subscriptions, donations, legacies, appeals, fund-raising and grants. Money raised is used for the conservation of habitats and species (partly through the purchase and management of land as nature reserves), research, education and for influencing land-use practices and government policies.

Local groups organise birdwatching trips, films, talks, fund-raising and promotional activities. Members can also get involved with practical conservation work on one of the reserves. For younger people the Young Ornithologists' Club (YOC) offers similar opportunities.

Membership facts

Number of members 680,000 (including 108,000 YOC members)

Membership fees Individual £12; joint £15; family £18; membership of the YOC costs £5 for one child, £6 covers all brothers and sisters

Magazine/newsletter *Birds* is a full-colour magazine published quarterly. YOC members' magazine is called *Bird Life* (partly printed on recycled paper)

Members also get Free admission to most RSPB reserves. Family members receive copies of *Birds* and *Bird Life*

Mail order A wide range of gifts is available through the mail order catalogue

Soil Association
86 Colston Street, Bristol BS1 5BB
BRISTOL (0272) 290661
With a goal of seeing 20 per cent of Britain's farmland in organic production by the year 2000, the Soil Association's aim is to promote the benefits of organic farming, humanely reared animals and chemical-free food. To achieve its aim, the association exposes the problems of using pesticides and other chemicals in food production and awards its symbol to foods, fertilisers and other goods that meet its standards for organic agriculture. To raise awareness of the issues, the Soil Association lobbies government, runs publicity

campaigns and publishes leaflets and educational material. Money to fund these activities is raised mainly through membership subscriptions and donations.

People who want to get more involved can also join one of 45 local groups.

Membership facts

Number of members   7000

Membership fees   Individual £12; family £18; students, unemployed and retired £8

Magazine/newsletter   *Living Earth*, published quarterly, is quite a weighty magazine of around 40 pages (the inside pages are recycled)

Members also get   Discounts on various products, updates on campaigns

Mail order   The association has a small mail order catalogue with books, T-shirts, etc.

The Tidy Britain Group (TBG)

The Pier, Wigan, Greater Manchester WN3 4EX

WIGAN (0942) 824620

The Tidy Britain Group campaigns for a litter-free, more beautiful Britain. The group is grant-aided by the Department of the Environment as the national agency for litter abatement. This money, plus money from sponsors, subscriptions and other initiatives, is used to fund its work of advising, educating and campaigning to prevent litter.

The group works with local councils, industries and communities to tackle litter problems.

The Tidy Britain Group publishes leaflets and advice packs with information about how people can help tidy up Britain – as individuals, groups or as companies. It also produces teaching materials for schools.

Individuals can join as associate members. The active involvement of members is welcomed to help with campaigning, local projects and other events.

Membership details

Number of members   500 (individuals)

Membership fees   Individual £10; family £15

Magazine/newsletter   Clean Nineties is a newsletter published quarterly on recycled paper

Members also get   Membership card, badge and sticker, plus a discount of 20 per cent on mail order goods

Mail order   Tidy Britain Enterprises produces T-shirts, books, badges, educational material and other goods

Transport 2000

Walkden House, 10 Melton Street, London NW1 2EJ

071-388 8386

Transport 2000 brings together many organisations concerned about the impact of transport on the environment and society. Transport 2000 believes that many of the problems – pollution, road congestion and accidents – could be solved by improving rail and bus networks and by making it easier and safer for people to walk or cycle.

Though membership is limited to organisations, Transport 2000 runs a supporters' scheme. By joining, you register your support and your money goes towards the organisation's work. If you want to get actively involved, you can join a local group and help raise money and awareness in your area. The groups also campaign on local issues.

Details of the supporters' scheme

It costs £20 a year to become a national supporter (£15 for the retired, unwaged and students, £25 for households). For that supporters receive ten issues of the newsletter *Transport Retort* and discounts on other publications and conferences. The fee to become an associate supporter is £6; associate supporters receive an annual report and cheap/free entrance to the AGM and summer conferences.

Women's Environmental Network

287 City Road, London EC1V 1LA

071-490 2511

The Women's Environmental Network is staffed entirely by women, and, although men are equally welcome to join, it takes a woman's perspective on environmental issues. Funded through subscriptions and donations, the network advises, informs, educates, campaigns and researches into topics that include the effect of chemicals in breast milk and food irradiation in pregnancy.

It promotes the concept of the 'green home' by showing how people can live in a way that is less harmful to the environment. It has been particularly active in campaigns to stop the use of chlorine-bleached pulp in products such as babies' nappies – lobbying government and manufacturers and getting involved in discussions with government departments over labelling.

Members are encouraged to attend meetings and seminars and to get involved with their local branch. Volunteers to help out in the office are always welcome.

Membership facts

Number of members 2000+

Membership fees Founder-member £20; ordinary member £10; unwaged £7

Magazine/newsletter A newsletter is sent out quarterly and *Green Living* magazine is published once a year. Both are printed on recycled paper

Members also get Information on campaigns. Founder-members get free entry to all conferences, seminars and meetings

Mail order None

World Wide Fund for Nature (WWF)

Panda House, Weyside Park, Godalming, Surrey GU7 1XR
GUILDFORD (0483) 426444

Part of an international organisation with offices in more than 20 countries WWF-UK has over 250,000 members. Money is raised through membership subscriptions, donations, legacies, appeals, from industry and through its mail order catalogue. Most of WWF's income is spent on

211

conservation projects and education. A third of the money raised in the UK is spent here, the rest used to fund projects overseas.

wwf's international work includes rainforest preservation, the conservation of habitats for endangered species like rhinos and the giant panda, and work to protect Antarctica. In the UK work includes habitat protection, campaigning to stop sea pollution and funding research into the effects of acid rain.

People who want to get actively involved can join a local supporters group (there are over 300 in Britain) and help with fund-raising and promotional activities.

Membership facts

Number of members   250,000

Membership fees   Ordinary £15; associate £20; fellow £30; companion £50; benefactor £100; junior £5

Magazine/newsletter   Published quarterly on chlorine-free paper, the newsletter *wwf News* is a tabloid-style paper printed in colour and sent to all members

Members also get   Associate members receive quarterly, and fellows monthly, copies of the bbc's *Wildlife* magazine in addition to *wwf News*. Companions and benefactors get the same benefits as fellow members. All members get a membership card and car sticker

Mail order   wwf's mail order catalogue sells a wide range of gifts, including writing paper, T-shirts and jewellery

Giving to charity

One way of 'investing' in a good cause is by making a donation to a charity that operates in the area you're particularly concerned about.

Covenants

Giving to charity via a covenant is a good idea if you're a taxpayer. A covenant is a legal document in which you promise to pay a set amount each year. To qualify for tax relief, the covenant has to be capable of running for at least four years. By using a covenant, every £75 you hand over is worth £100 to the charity, as it can claim back basic-rate tax. If you're not a taxpayer, making payments to a charity through a covenant isn't worth it, as you would have to pay the Revenue any tax deducted from what you give.

You usually don't have to worry about drawing up complicated legal documents – most charities will send a printed form and arrange for a standing order. You may also be able to pay one lump sum to cover the number of years you intend to donate for – called a 'deposited covenant'.

Charities Aid Foundation

You don't have to commit yourself to the same charity for four years. Instead, you can make out a covenant to the Charities Aid Foundation (CAF) – a special charity that makes donations to other charities under your instructions.

The money you give, plus the tax credits, is kept in a special account, and CAF issues you with a book of charity vouchers. These work like cheques – you fill one in with the amount you want to give, send it off to your chosen charities and they then claim the money from CAF. If you prefer, payment can be made direct from CAF on your instruction. CAF can split your donation between as many charities as you wish. CAF's address is on page 217.

Legacies and gifts

Outright gifts to registered charities are not liable for inheritance tax, and there is no tax to pay if you leave assets to registered charities in your will.

Give-as-you-earn

The official name is the Payroll Deduction Scheme and, in the 1990–91 tax year, it allows employees (if their employer offers the scheme) to give up to £50 a month or £600 a year to charity and get full tax relief on the contributions. So for a basic-rate taxpayer a donation of £10 really costs only £7.50. You can nominate any organisation from a wide range of categories recognised by the Revenue, which may not necessarily be registered charities – hospitals and play-groups, for example, may qualify.

The Inland Revenue publishes two useful free leaflets: IR64 Giving to charity – how businesses can get tax relief, and IR65 Giving to charity – how individuals can get tax relief.

Gift Aid

Introduced in the 1990 Budget, this scheme allows individuals tax relief on single gifts to charity of not less than £600 and not more than £5 million. The donor makes the gift net of basic-rate tax and the charity reclaims this from the Inland Revenue. Higher-rate tax relief will also be available to relevant donors. Gift Aid applies to donations made from 1 October 1990.

Affinity cards

These are credit cards that are linked to specific charities. The idea is that a percentage of what you spend using the card goes to the charity. Some also donate a fixed sum (usually around £5) to the charity when you first join. Companies offering such cards and the charities they're linked to include:

Bank of Scotland Visa – NSPCC; Co-op Visa – RSPB, Help the Aged; Girobank Visa – Oxfam; Leeds Permanent Building Society Visa – British Heart Foundation, Imperial Cancer Research Fund, MENCAP; Midland Bank Care Card – Age Concern, British Diabetic Association, British Red Cross, Cancer Research Campaign, Multiple Sclerosis Society, NSPCC, RNIB, RNID, Save the Children Fund, Spastics

Society, St John Ambulance, Terrence Higgins Trust; Natwest World Wide Fund Visa – World Wide Fund for Nature; Royal Bank of Scotland Visa – Royal National Lifeboat Institution, Woodland Trust, Canine Defence League.

USEFUL ADDRESSES

British Board of Agrément (BBA)
PO Box 195
Bucknalls Lane
Garston
Watford
Hertfordshire WD2 7NG
WATFORD (0923) 670844

British Glass Manufacturers Confederation
Northumberland Road
Sheffield S10 2AU
SHEFFIELD (0742) 686201

British Plastics Federation
5 Belgrave Square
London SW1X 8PO
071-235 9483

British Waste Paper Association
Alexander House Business Centre
Station Road
Aldershot
Hampshire GU11 1BQ

Can Makers Information Service
36 Grosvenor Gardens
London SW1W 0ED

Charities Aid Foundation
48 Pembury Road
Tonbridge
Kent TN9 2JD
TONBRIDGE (0732) 771333

The Community Furniture Network
Highbank
Halton Street
Hyde
Cheshire SK14 2NY
061-367 8780

Confederation for the Registration of Gas Installers
  (CORGI)
St Martins House
140 Tottenham Court Road
London W1P 9LN
071-387 9185

Department of Transport
VCA Division
Room 805
Tollgate House
Houlton Street
Bristol BS2 9DJ

Heating and Ventilating Contractors' Association
34 Palace Court
Bayswater
London W2 4JG
071-229 2488

Institute of Plumbing
64 Station Lane
Hornchurch
Essex RM12 6NB
HORNCHURCH (040 24) 72791

217

London Waste Regulation Authority
The County Hall
London SE1 7PB
071-633 4221

National Association of Plumbing, Heating and Mechanical Service Contractors
Ensign House
Ensign Business Centre
Westwood Way
Coventry
West Midlands CV4 8JA
COVENTRY (0203) 470626

Nature Conservancy Council
Northminster House
Peterborough
Cambridgeshire PE1 1UA

The Royal Society for Nature Conservation (RSNC)
see page 206

The Royal Society for the Protection of Birds (RSPB)
see page 207

Scottish and Northern Ireland Plumbing Employers Federation
2 Walker Street
Edinburgh EH3 7LB
031-225 2255

Solid Fuel Advisory Service
Hobart House
Grosvenor Place
London SW1X 7AE
071-235 2020

Waste Watch
26 Bedford Square
London WC1B 3HU

Working Weekends on Organic Farms
19 Bradford Road
Lewes
East Sussex BN17 1RB